GO TO CUSTOMER

GO TO CUSTOMER

A Guidebook to Selling What's Possible™

– for Enterprise Account Teams

Dave Irwin

POLARIS I/O
Chicago, IL

Published by Polaris I/O
Chicago, Illinois
polarisio.com
info@polarisio.com

Polaris I/O books are available at special quantity discounts for bulk purchase for sales promotions, events, fundraising, and educational needs. Special books or book excerpts also can be created to fit specific needs. For details and permission requests, write to the email address above.

ISBN 979-8-9932719-1-0 (eBook)
ISBN 979-8-9932719-0-3 (paperback)
ISBN 979-8-9932719-2-7 (hardback)
ISBN 979-8-9932719-3-4 (audiobook)

10 9 8 7 6 5 4 3 2 1

—

Copyediting by James Gallagher
Graphics by Tim Rawls
Proofreading by Adeline Hull
Book Design & Publishing by Kory Kirby
SET IN MINION PRO

CONTENTS

Introduction   *vii*

CHAPTER 1:
Unlocking Your Invisible Pipeline   **1**

CHAPTER 2:
Every Insight Needs a Story   **25**

CHAPTER 3:
Understanding the Customer's Business Environment   **55**

CHAPTER 4:
Seeing Around Corners Inside Your Account   **67**

CHAPTER 5:
Making Connections   **107**

CHAPTER 6:
Selling What's Possible   **121**

CHAPTER 7:

Planning for Broad and Deep Relationships   **145**

CHAPTER 8:

What Customers Value   **171**

CHAPTER 9:

Who Customers Value   **197**

CHAPTER 10:

Communicating Value   **219**

CHAPTER 11:

Building a Commercial Community   **233**

AFTERWORD:

It's Time To Move   **249**

Appendix 1   *251*

Acknowledgments   *263*

About the Author   *265*

INTRODUCTION

If you lead or work on an enterprise account team, you already know how overwhelming the job has become. Every day, you juggle shifting customer needs, competing internal priorities, spreadsheets, PowerPoint presentations, customer relationship management (CRM) updates, and endless meetings. And even after all that effort, you still end up blindsided.

One day you think you're aligned with your customer, the next you find out their priorities have shifted without warning—and suddenly your plan, your pitch, and sometimes even your relationship is outdated.

It's not your fault. The tools and systems you're given aren't built to help you see your customer clearly. CRM systems are designed to track what you want to sell, not what your customer needs. Account reviews are quarterly, backward-looking, and often more about what your team hopes to push into the market than about what customers are investing in right now. Spreadsheets and meetings try to fill the gap, but they can't keep pace with the speed at which customers' needs change. The result is wasted time, missed opportunities, and, too often, the sinking feeling of being out of step with the very people you're supposed to be serving.

And the complexity keeps multiplying. COVID didn't just disrupt work—it rewired how decisions are made, how people collaborate, and even how much time stakeholders have to give you. Political upheaval, digital transformation, and the rise of artificial intelligence (AI) have only accelerated the uncertainty. Analyst firms now report what you already feel: Burnout is high, goals aren't being met, and most of the tools out there don't make your job any easier.

But you also know what great looks like. You've seen moments where it all clicks—where your team is in rhythm with stakeholders, where you're solving problems side by side, where the relationship feels less like supplier and customer and more like two teams innovating together. Those are the moments that make the work meaningful. The question is: Why are they so rare? Why is it so hard to make that kind of success repeatable? And what would it mean if you could scale those moments across every account?

That's where this book comes in. *Go to Customer* lays out a new strategy for account growth—one designed for today's reality. It's a strategy that reverses the old flow of pushing products onto customers and instead aligns your value to what customers actually need and are already investing in. Instead of chasing leads, you'll learn how to uncover needs as they emerge and match them with relevant solutions in the exact moment they arise. Instead of pushing products, you'll learn how to partner with customers to create what's possible—working backward from their priorities and turning those priorities into pipelines of opportunity. The promise? More relevance with customers, less wasted time, faster growth, and stronger, more resilient relationships.

Today the stakes are clear:

If you aren't growing an account, you're losing it.

Customers no longer stay loyal to suppliers by default. Buyers are more empowered than ever, able to access their own information and make their own decisions. The shift in wallet share will be faster and more unforgiving than ever before. But one constant remains: The world's biggest companies control the majority of enterprise spending—and unlocking growth within them is the surest path to success.

And how do you unlock growth? Make what the customer needs, and is spending money on, the cornerstone of any account plan and expansion strategy. A.k.a. *Go to Customer.*

If you manage or run account teams, this book is designed to help you break free from the grind of chasing what customers might want—and instead guide them where they want to go. You begin with insights, progress through planning and execution, and then manage the customer experience digitally with personalized online interactions tailored to each stakeholder. This is the process we will walk through in this guidebook step-by-step. By the end, you'll have a repeatable, scalable system to grow enterprise accounts, strengthen customer relationships, and open up new worlds of opportunity.

HERE'S HOW THIS GUIDEBOOK IS STRUCTURED AND WHAT'S AHEAD FOR YOU

- **Chapter 1: Unlocking Your Invisible Pipeline.** You'll learn how to uncover the vast, hidden opportunities inside your existing accounts and why customer-driven needs are the most powerful source of growth.
- **Chapter 2: Every Insight Needs a Story.** You'll see how to turn raw insight into compelling narratives that win the attention, trust, and budgets of customer stakeholders.
- **Chapter 3: Understanding the Customer's Business Environment.** You'll gain tools to navigate volatility, uncertainty, complexity, and ambiguity (VUCA) so you can stay aligned even as customer conditions shift.
- **Chapter 4: Seeing Around Corners Inside of Your Account.** You'll discover how to anticipate risks, changes, and disruptions inside your accounts before they blindside you.
- **Chapter 5: Making Connections.** You'll learn how to map influence networks, build broader coalitions of support, and become a hub for customer collaboration.
- **Chapter 6: Selling What's Possible.** You'll uncover how to grow deal sizes by engaging at the right level and aligning to bigger-picture outcomes.
- **Chapter 7: Planning for Broad and Deep Relationships.** You'll learn how to modernize account planning with dynamic, daily intelligence rather than static quarterly reviews.
- **Chapter 8: What Customers Value.** You'll see how to align solutions directly to outcomes customers are investing in, rather than to your own product goals.

- **Chapter 9: Who Customers Value.** You'll understand the attributes customers look for in the people they trust, and how to structure account teams around those expectations.
- **Chapter 10: Communicating Value.** You'll master how to frame proposals, messages, and stories that resonate with the decision-makers who matter most.
- **Chapter 11: Building a Commercial Community.** You'll learn how to extend your influence by building connected ecosystems around your customers and their priorities.

Each chapter builds on the last, giving you the mindset and practical tools to make customer alignment systematic and repeatable.

But before we get to all of this, perhaps you're wondering . . .

HOW DID I ARRIVE AT THIS GO TO CUSTOMER STRATEGY BEING THE KEY TO ENTERPRISE ACCOUNT SUCCESS?

Well, I've spent over three decades managing enterprise accounts. I've served as chief marketing officer (CMO), chief revenue officer (CRO), president, head of strategy, and division leader across leading data, analytics, and AI companies. I was in enablement before it was even called enablement. I've led account teams, mapped relationships, driven expansion strategies, built aligned products and solutions tied to business impact, and designed systems for growth. Along the way, I've trained hundreds of account sellers and sales teams, mapped customer problems and needs, and built large-scale account structures to make it easier for sellers to communicate value in solving customer problems. I've felt the frustrations, made hundreds of mistakes, and celebrated the breakthroughs.

I began to observe that there wasn't an effective system to help account teams truly know their customers. So I created one—because what drives is a passion to make this process easier, more systematic, and more rewarding for account teams and customers alike.

My experiences have granted me a unique vantage point to what consistently works and what quietly wastes time inside enterprise account teams. Across all the years, all the cycles, and all the changes . . . one pattern kept repeating: Account teams hit ceilings when they lead with what they are told to sell instead of what customers are already mobilizing to buy.

When teams push what they are told to sell, growth stalls. When they work in rhythm with enterprise stakeholders—align people, stories, and capabilities to outcomes that matter inside the account—everything changes. Trust forms. Walls drop. It stops being supplier versus customer and becomes connected teams solving problems and innovating together. When you operate that way, relevance rises, friction falls, and growth becomes repeatable. These results are why I codified *Go to Customer* as a practical system you can run every day, step-by-step.

This strategy is built on decades of research and has been applied by teams who have systematically won hundreds of contracts and bigger, more profitable deals with shorter sales cycles, bigger average deal sizes, and large-scale pipelines to prioritize and pursue. One time I watched a single, well-equipped account team—led by a top seller and supported by experts—generate twenty times more in contract value for one client, at a fraction of the cost of a much larger sales team that was focused on new-prospect sales. The difference wasn't heroics; it was a systematic, insight-driven approach.

So the choice is to stick with outdated product centric approaches or embrace a customer-centric, data-informed strategy that uncovers hidden value and drives lasting growth.

My goal in writing *Go to Customer* was simple: Make every account team successful by removing today's obstacles and opening a new world of growth.

Let's unlock what's possible together.

CHAPTER 1:

UNLOCKING YOUR INVISIBLE PIPELINE

The room was quiet, punctuated only by the soft tapping of laptop keys, the nervous shuffling of papers, and the occasional clinking of a coffee cup. At the head of the table, I posed a simple question: "What is your enterprise growth strategy for this account?" Faces stiffened. Some glanced at their notes; others avoided eye contact. Silence lingered, heavy with uncertainty.

In that moment it became clear—no one had a confident answer. This talented, hardworking team was focused on what they knew: current contracts, upcoming renewals, and a handful of established contacts. Beyond that familiar territory, the possibilities were a blank map.

RELENTLESS GROWTH STARTS HERE: WHY IGNORING CUSTOMER NEEDS IS COSTLY

Account teams and the companies they work for often don't realize the massive opportunity for cross-selling and upselling, given the current levels of penetration within enterprise accounts. Imagine having one to two hundred more opportunities available to you each year, perhaps ten to twenty times greater than current pipelines. What if these opportunities were two to three times the average deal size and closed at a much higher rate, since they originate from the customer's need to solve a problem? Companies are leaving hundreds of millions in potential revenue on the table without even realizing it.

Now think about the cost of account teams spending time on the wrong opportunities or sitting idle, perhaps doing a lot of internal tactical work instead of pursuing these larger opportunities. This represents a staggering waste of time for customer-facing resources. You wouldn't accept a factory worker who spends 70 percent of their time off the manufacturing line, but we willingly accept this with our account teams. Why? Because we're not really sure what they are doing. If they're not pursuing much larger pipelines, they're likely focused on internal or low-value work. I've seen this scenario play out hundreds of times. Teams are diligent about managing the business they already have, but rarely do they venture beyond it.

The idea of exploring a client's vast, multilayered organization, finding new divisions, new leaders, and new needs, is overwhelming. After all, most account teams have never been given the tools or training to dig deep, to research, and to analyze the full scope of what's possible within a massive enterprise. Have you ever seen an account executive job description that lists extensive research and analytical skills as a requirement? I haven't.

Often the real culprit is hidden in plain sight: the way companies organize themselves. By dividing teams by product lines and assigning quotas to specific territories, organizations unintentionally encourage a narrow focus. Account teams become experts in their own corners but remain unaware of the bigger picture. The customer's world—dynamic, complex, and full of potential—goes largely unexplored.

That moment of hesitation, that silent pause, is more than just a knowledge gap. It's a missed opportunity repeated across countless teams and accounts simply because no one has provided them the capability to see the whole enterprise, not just the piece they already know.

FROM VENDOR TO TRUSTED ADVISER: MASTERING THE ART OF CUSTOMER-CENTRIC GROWTH

Navigating the Customer's World: A Call for Insight and Understanding

Today's customers are navigating a complex landscape of priorities and challenges, many of which extend far beyond the scope of any single vendor's products or services.

> *Research indicates that only 11 percent of sales conversations are perceived as valuable by customers. Even more striking, Forrester reports that 81 percent of buyers remain dissatisfied with their experience, even when they ultimately select a vendor.[1]*

---

[1] Forrester, "Forrester: To Master B2B Buying Mayhem, Providers Must Prioritize Buyers' Needs," *Forrester* (press release), December 4, 2024, https://www.forrester.com/press-newsroom/forrester-the-state-of-business-buying-2024/.

This disconnect is not new. Executive buyers consistently express frustration that account teams lack a deep understanding of their business environment and unique roles. As a result, sellers frequently fail to deliver meaningful insights or innovative ideas that customers need to achieve their objectives.

Imagine, for a moment, sitting in a customer's chair. The reality faced by customers is often far more complex and demanding than many account teams realize. One account leader candidly shared that her primary responsibility is to shield her customer stakeholders from her own product sellers because the interactions can be so disruptive and unhelpful that she is asked to keep them away. In extreme cases, customers have even gone so far as to ban certain supplier individuals from contacting them. This is the current reality, and executives often are not aware that this is happening.

This should serve as a call for change. Every customer interaction should be approached with curiosity, empathy, and a commitment to understanding the customer's unique challenges. By doing so, the account team not only differentiates itself from the competition but also builds the trust and credibility necessary for long-term partnership and success.

Transform Customer Engagement: Insightful, Empathetic, and Impactful

Customers are not simply seeking products or services. They are looking for genuine partners who demonstrate sincerity, empathy, and a true commitment to solving their pressing challenges. It is essential to recognize that every customer initiative, especially those that secure funding, is driven by a critical internal or external event. These events, whether they involve launching new products, entering new markets, navigating organizational changes, implementing technology upgrades, pursuing digital transformation, or managing risks, are constantly unfolding across departments and business units, often globally.

As these initiatives cascade throughout the organization, they become the focal point of customers' investments in time and resources. It is crucial, therefore, to approach every interaction with a deep understanding of the customer's current priorities and threats, pain points, and aspirations. By providing actionable insights and tailored solutions, one is positioned as a trusted adviser rather than just another vendor.

Customer Triggered Growth Opportunities

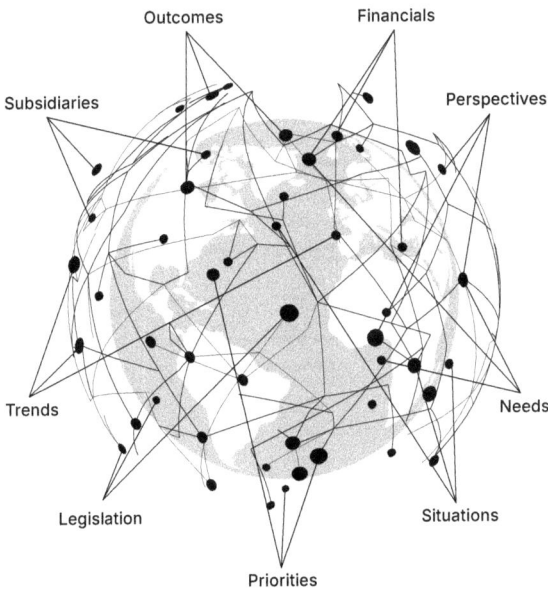

1.1: Customer Needs and Challenges are Many

Having established why a trusted partnership matters, the next step is to unlock and nurture sustainable demand within those relationships through proven frameworks.

Ignite Account Growth: Reveal, Activate, and Accelerate Hidden Opportunities

Accelerating cross-sell and upsell revenue is based on addressing customer challenges and strategic initiatives. Engaging early with high-value accounts is transformative to profitable growth. These accounts often contain untapped, customer-driven opportunities—an "invisible pipeline" of growth potential. Too often, this is missed due to manual processes, information overload, and a lack of centralized knowledge sharing.

Common Research Challenges Stand in the Way of Understanding Customer Needs

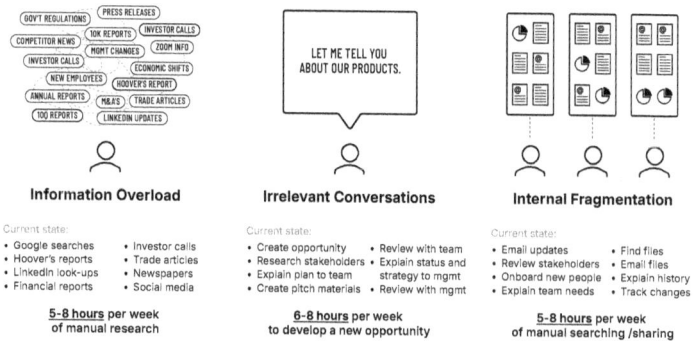

Information Overload

Current state:
- Google searches
- Hoover's reports
- LinkedIn look-ups
- Financial reports
- Investor calls
- Trade articles
- Newspapers
- Social media

5-8 hours per week of manual research

Irrelevant Conversations

Current state:
- Create opportunity
- Research stakeholders
- Explain plan to team
- Create pitch materials
- Review with team
- Explain status and strategy to mgmt
- Review with mgmt

6-8 hours per week to develop a new opportunity

Internal Fragmentation

Current state:
- Email updates
- Review stakeholders
- Onboard new people
- Explain team needs
- Find files
- Email files
- Explain history
- Track changes

5-8 hours per week of manual searching /sharing

1.2: Common Research Challenges.

Teams today tend to focus on existing contracts and renewals, limiting their external view of the customer. The root cause is a lack of visibility into evolving customer priorities—information that rarely appears in traditional account plans because it's hard to see. Occasional announcements or articles are shared, but these rarely provide the depth needed for meaningful action and engagement.

Furthermore, when probing for relationships and understanding of key decision-makers, many influential executives remain unknown to the account team. While enterprise communication plans are sometimes developed, they frequently lack actionable insights into what these stakeholders actually value. Relying on teams to simply "figure it out" is neither efficient nor effective.

The Evidence Is Clear: Engage Early

When account teams engage customers at the inception of funded initiatives—when business priorities are being set—they are better positioned to align with larger budgets and strategic objectives. Early involvement enables teams to become trusted advisers, offering solutions and insights that resonate with stakeholders' real concerns.

To excel in today's competitive environment, it is essential to be present, relevant, and responsive to the customer's situation in the context of what is occurring. Building credibility by demonstrating genuine interest in solving customer problems has become essential to success. By embedding your account team in the customer's journey and consistently providing timely, contextually relevant insights, you not only win business; you build lasting partnerships that fuel sustained growth. The imperative is clear: Move beyond only what you see through daily interactions, embrace an insight-driven customer-centric approach, and unlock the full potential of every account.

How Funded Initiatives Are Born

1.3: Creating Demand—Engage Early When Customer Needs First Arise

STOP CHASING, START CULTIVATING: THE PROVEN PATH TO GROWTH LIES WITHIN YOUR EXISTING CUSTOMERS

Transform Pipeline Development: From Passive Leads to Active Needs

Imagine pipeline development not as a hunt for scattered prospects, but as the discovery of an untapped gold mine within your existing account base. Traditionally, organizations have focused their investments on generating new sales leads—chasing transactional opportunities rather than nurturing strategic relationships. This approach often overlooks the immense potential for upselling and cross-selling within key accounts.

Transform Pipeline Development

1.4: Rethinking Pipeline Development

Consider this:

> Gartner reports that 58 percent of organizations fail to
> meet their quotas for key accounts.[2]

This statistic is a clear warning that current strategies are not enough.

Shifting from Lead Gen to Need Gen

The real opportunity lies in shifting from a lead generation mindset to a *need generation* mindset. By understanding and addressing the evolving needs of enterprise customers, especially those in the Global 2000, organizations can unlock a vast, invisible pipeline of growth. These customers represent entire markets, each with unique challenges and opportunities to address with your help.

---

2 "Revamp Your Key Account Management Strategy: Break the Trend of Underperformance and Unlock Growth," *Gartner*, accessed September 21, 2025, https://www.gartner.com/en/sales/trends/key-account-management-strategy.

Now is the time to pivot. Instead of merely seeking new leads, invest in uncovering and fulfilling the needs of your most valuable customers. Engage with them, listen to their challenges, and collaborate on tailored solutions. This proactive approach not only strengthens relationships but also drives sustainable, long-term growth. The gold mine is there waiting for those ready to mine it.

Total Addressable Account Pipeline

Unmet customer needs provide
a large additional pipeline

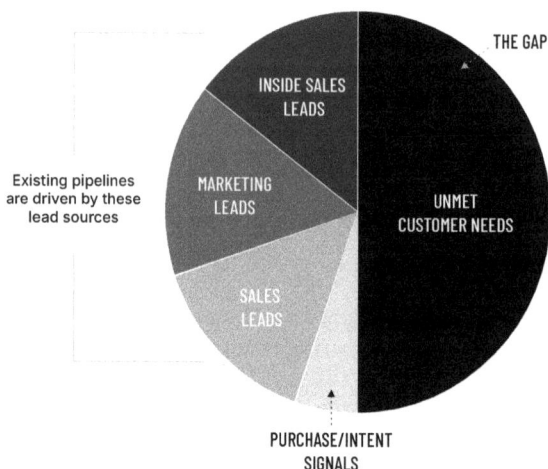

THE GAP

INSIDE SALES
LEADS

Existing pipelines
are driven by these
lead sources

MARKETING
LEADS

UNMET
CUSTOMER NEEDS

SALES
LEADS

PURCHASE/INTENT
SIGNALS

Unmet Customer Needs
The missing piece

- **Expansive source** of upstream customer expressed needs
- **10-20x volume** of typical average CRM pipeline
- Translate into **larger customer funded strategic initiatives**

- **Span departments, subsidiaries, geographies, and levels**
- **Competitive advantage** by engaging emerging clients early
- Proactive insights **unlock significant growth opportunities**

1.5: Discover an Untapped Pipeline

TRANSFORM ACCOUNT GROWTH BY SOLVING REAL CUSTOMER PROBLEMS, NOT JUST SELLING MORE PRODUCTS

From Product Selling to Problem-Solving

By deeply understanding and addressing customer needs, you create meaningful value and sustainable growth. This mindset transforms your role from a vendor to a trusted partner, cultivating long-term success across the entire customer footprint.

CLIENT NEED   DESIRED STATE

CURRENT STATE
From Product Selling
Most account sellers focus on pitching product features and functions—a strategy that no longer works with today's empowered buyers. Consider this: 81% of buyers are unhappy with their experience (Forrester), 59% of sellers fail to grasp buyers' goals (Salesforce), and 60% of qualified deals end with no decision made (Corporate Visions).

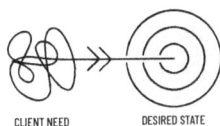

FUTURE STATE
To Problem Solving
When account sellers get in early—before executive buyers have started addressing a core challenge—they can create, shape, and win new deals. Win rates double, deal sizes grow by 2-3x, and sales cycles shorten, with fewer deals ending in "no decision." Account sellers who co-create solutions with customers win bigger deals faster.

1.6: Transform Account Growth—from Product Selling to Problem-Solving

White Space vs. Bluespace™: Rethinking Account Growth for Lasting Customer Relevance

Traditional white space analysis evaluates which products have or haven't been sold to various business units. However, this inside-out approach assumes your products align with different areas of your customers' business rather than with all the needs customers have that could be addressed by your company's overall capabilities.

In contrast, a Bluespace analysis focuses on the customer's real, timely challenges and needs. Here the analysis is reimagined: Customer problems form one axis, while buying centers, across departments and business units, form the other. This outside-in approach uncovers genuine opportunities for partnership and value creation.

Embracing the Bluespace Approach

To unlock sustainable, meaningful account growth, organizations must shift to a Bluespace approach. Unlike white space, Bluespace analysis adopts an outside-in view—focusing not on what you want to sell, but on the real problems your customers need to address.

Our experience shows that customers consistently articulate their most pressing challenges, both in direct conversations and public forums. Their investments follow these priorities, with budgets allocated to initiatives that address urgent needs. Critically, the decision-makers and influencers behind these initiatives often span traditional organizational boundaries, forming dynamic buying networks that require a more holistic, adaptive approach.

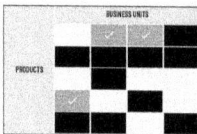

CURRENT STATE
From Whitespace
Whitespace, designed to look for new areas to sell products by business unit, is the opposite of what executive buyers need. Executive buyers don't care about products and disengage quickly with account sellers.

FUTURE STATE
To Bluespace™
Bluespace looks at areas of opportunity from the buyer's perspective, based on problem categories by buying center—often involving stakeholders who span multiple business units. Bluespace aligns with buyer needs.

1.7: Embracing the Bluespace™ Mindset—from White Space to Bluespace™

Actionable Insights and Cautionary Guidance

Aligning your solutions with the specific, timely challenges your customers face and engaging the right stakeholders as those needs emerge positions your organization as a trusted partner, not just another vendor. This approach demands continuous listening, agility, and a willingness to move beyond product-centric thinking. Beware of defaulting to internal goals; instead, let customer relevance guide your strategy.

Motivation for Change: Elevating Opportunity and Impact

Today's Global 2000 enterprises are sprawling ecosystems, with hundreds or even thousands of subsidiaries and business units, each confronting unique challenges every year. By systematically cataloging customer problems and mapping them to the right buying centers, especially at the critical moments when funded initiatives are planned or launched, you ensure your offerings are visible, timely, and directly relevant.

In an increasingly competitive market, relevance is no longer optional. It is essential. Adopting a Bluespace approach empowers your teams to lead with intelligence, act with precision, and build resilient, trust-based customer relationships. Challenge your teams to look beyond the grid: Focus on the problems that matter, at the moments they matter most, and with the people who can drive real change.

Embrace this shift. Transform your sales strategy and customer success by becoming the indispensable resource your customers rely on to solve their most pressing challenges.

Turn Customer Insight into Powerful Pipeline Momentum: Make Growth Systematic and Real

Adopting a customer problem-centric perspective, rather than a purely product-focused one, is highly effective for enhancing the quality of your growth pipeline. High-performing pipelines are built on a thorough understanding of the diverse challenges our customers face—both immediate and long term, recurring and evolving, established and emerging. These problems may vary by geography, business unit, or organizational level, and recognizing these distinctions is critical.

To maximize impact, you should consistently analyze the full spectrum of customer issues. This enables you to strategically select where to invest your resources for the greatest return, while maintaining a holistic view of the business landscape. Remember, balancing the quantity of opportunities with their quality is key. Focus on cultivating opportunities that offer sustained value, ensuring your growth compounds year after year.

By anchoring your strategy in the real, evolving needs of your customers, you not only unlock greater long-term revenue but also deepen and future-proof your client relationships. Stay vigilant, stay curious, and let customer problems guide your path to enduring success.

Pipeline Quality

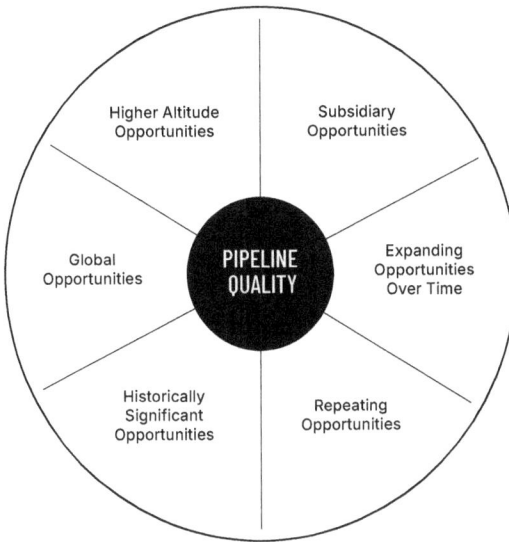

Higher Altitude
Opportunities

Subsidiary
Opportunities

Global
Opportunities

**PIPELINE
QUALITY**

Expanding
Opportunities
Over Time

Historically
Significant
Opportunities

Repeating
Opportunities

1.8: Dimensions of Pipeline Quality

ANALYZING AND SYNTHESIZING CUSTOMER NEEDS: THE POWER OF A GROWTH MINDSET

To truly understand and address customer needs, organizations must adopt a **growth mindset**—an outlook that embraces different possibilities, challenges assumptions, and encourages creative problem-solving beyond the limitations of existing products. This mindset is not just aspirational; it is foundational for building trust, fostering collaboration, and cocreating solutions that deliver real value and inspire buyer confidence.

Through experience, we have identified **seven critical mindset shifts** essential for anyone seeking to break free from a fixed, product-centric approach and move toward a solution-oriented, customer-first perspective. These shifts cultivate self-awareness, reduce bias, and enable effective engagement with customers as partners in innovation.

Seven Mindset Shifts for Customer-Centric Problem-Solving

1. **Visualize Current and Future States:** Use visual tools to map both the present situation and the desired outcomes. Illustrating obstacles and opportunities helps clarify the path forward and makes complex issues more accessible for customers. Approach each scenario as a detective would, uncovering inefficiencies and areas for improvement, then collaboratively design solutions.

2. **Cultivate Curiosity:** Let curiosity drive your inquiry into the root causes of customer challenges. Resist the urge to default to product pitches; instead, observe, question, and break down the root causes of each problem or the gap between the current and desired state.

3. **Adopt Multiple Perspectives:** Recognize that every stakeholder views a problem differently depending on their role and vantage point. Step back for a big-picture view, then zoom in on critical details. This layered approach uncovers opportunities that might otherwise remain hidden.

4. **Lead with Empathy:** Put yourself in your customer's position. Understand their pressures and constraints. Demonstrate genuine concern by reflecting their challenges back to them and validating their experiences. Empathy builds trust and alignment, bridging the gap between your solutions and their needs.

5. **Pursue Thoughtful Analysis:** Avoid rushing to conclusions. Take the time to analyze problems thoroughly, identifying sustainable improvement opportunities. Careful, deliberate thinking uncovers details that can be pivotal in securing customer buy-in.

6. **Be Action Oriented:** Move beyond recommendations—initiate ongoing dialogue and provide actionable insights into how a recommendation should be implemented and why specific actions will yield the desired outcomes. Consistent engagement positions your team as proactive partners, not just vendors.

7. **Act as a Trusted Guide:** Lead customers confidently through the solution process. Provide clear road maps, resource plans, risk assessments, and success metrics. Proactively address doubts to prevent delays and build consensus among stakeholders.

UNLOCK EXPONENTIAL GROWTH: MASTER A PROVEN GROWTH FRAMEWORK

Energize Opportunity Discovery: Structured, Strategic, Unrivaled

Identifying and developing new opportunities requires both creativity and discipline. Years of experience have shown that a **repeatable framework** enables teams to consistently uncover, evaluate, and pursue high-value opportunities. Given the significant investment required, often in the six-figure range, it is imperative to approach each opportunity with rigor and clarity.

A Framework for Cultivating Actionable Growth Opportunities

- **Challenge:** Clearly articulate the core issue from the customer's perspective, including context and impact.
- **Situation:** Detail the origin, stakeholders involved, and the significance of the problem.
- **Buying Center Stakeholders:** Map out all affected parties, their relationships, and their desired outcomes. Consider each perspective to ensure alignment and consensus.
- **Key Roles:**
  - *Hero:* The champion driving the vision.
  - *Guide:* The adviser navigating internal dynamics.
  - *Wallet Owner:* The decision-maker controlling budget approval.
  - Maintain regular communication and tailor input for each VIP to build trust and momentum.
- **Desired Outcome:** Define a shared vision, success metrics, and the anticipated business impact. Ensure all stakeholder perspectives are integrated.
- **Solution:** Present a compelling future state, supported by detailed use cases, components, and a clear implementation road map that is foundational to establishing a shared vision of success with important stakeholders.
- **Value Statements:** Craft concise, targeted value propositions for each core stakeholder, highlighting your unique differentiators.
- **Execution Team:** Identify internal team members, assign clear responsibilities, and demonstrate alignment with customer stakeholders.

- **Customer Intent:** Capture evidence that your customer intends to move forward and reinforce this intent across their buying network.

This disciplined process ensures clarity, reduces risk, and maximizes the likelihood of success. Remember: **Slowing down to analyze and plan is the fastest route to sustainable growth.**

The Emerging Role of the Commercial Insight Strategist

We've introduced a new role for unlocking the invisible pipeline and breaking down opportunities into structured actionable reports that connect a customer's challenge or need, the stakeholders involved, the situation, desired outcomes, potential solutions, and value propositions, as well as keeping these updated. We call this new role the commercial insight strategist (CIS). Like a business development rep (BDR), whose job is to get prospecting meetings with leads, a CIS helps account leaders get customer meetings based on needs. This teammate to the account executive is an insight specialist who analyzes account intelligence, organizes opportunities, and helps account executives prioritize and engage effectively. It's an essential role that we reference throughout the book and profile in more depth in chapter 9.

The CIS role will help ensure that account executives are fed important urgent account opportunities based on needs and challenges daily. Like a BDR, they are an essential part of the workflow in identifying, cultivating, and winning new insight driven opportunities working backward from customers by connecting all the dots for the account executives, keeping them informed along the way, and potentially even engaging with account stakeholders directly with account executives.

Deal Pursuits: Protecting Significant Investments

Pursuing large deals is a major investment often exceeding hundreds of thousands of dollars when accounting for the collective time of all involved, including meetings and documentation. Yet many organizations fail to systematically protect this investment, risking loss of knowledge and momentum. It is essential to centrally manage assets, learning opportunities, and tribal knowledge, and to assign clear ownership throughout the pursuit.

Embracing Pipeline as a Service: A New Workflow to Streamline Account Growth

> *Traditional account teams are often overwhelmed by administrative tasks, with more than half of their time spent on non-revenue-generating activities.*[3]

To unlock efficiency and growth, organizations should adopt a workflow oriented approach that supports pipeline development as a service:

- **Upskill** roles like sales development representatives (SDRs) and BDRs to become CISs, focused on identifying and packaging customer needs for account teams. For a deeper understanding of the origins and critical importance of this role in achieving success within today's insight-driven account growth, see chapter 9.
- **Systematize** discovery and opportunity development to

---

3 Strategic Account Management Association (SAMA), https://strategicaccounts.org.

enable account sellers to repurpose up to 50 percent of their time for high-value customer engagement activities.

- **Reverse the flow** by taking an outside-in view of customer needs. In this way, account teams can uncover new, high-potential opportunities, access larger budgets, and position themselves strategically.

Organize your pipeline using multiple lenses—geography, subsidiary, business impact, and problem type—to reveal untapped growth areas and optimize resource allocation.

Balancing Logic and Emotion in Decision-Making

While logical alignment with customer needs is vital, it is important to recognize that executive buyers also make decisions emotionally. Storytelling should be leveraged to connect both rational and emotional drivers, ensuring that solutions resonate on every level. Refer to chapter 2 for more on storytelling.

Proactive Demand Creation: Your Competitive Advantage

Creating demand is far more effective than merely responding to it. By shifting focus from leads to needs, you open new avenues for value creation and revenue growth. Start small—pilot working backwards from customer needs in a systematic way with one or two accounts. Rapid results are achievable, as evidenced by teams who have seen transformative outcomes in just six weeks.

Take Action: Build Your Growth Pipeline as an AI-Enabled Service Workflow

Utilize the provided template and checklist to launch your need-based pipeline development workflow today. This structured approach is foundational to exponential pipeline growth, enhanced client value, and increased productivity for your account teams. The future of customer engagement and growth is here—embrace it with confidence and discipline.

ROLE BY ROLE: STEPS TO SHAPE THE FUTURE

For Account Teams

A steady influx of pipeline opportunities is invaluable, but true success hinges on the relevance of these opportunities to your customers' unique priorities. When opportunities originate directly from your customers' areas of interest and investment, your win rates and deal sizes consistently improve. The most effective approach is to engage early and collaboratively craft a compelling narrative that demonstrates how you and your customer can solve critical business challenges together. Remember, relevance drives results.

For Marketing

Embracing both "need generation" and "lead generation" strategies unlocks a new dimension of pipeline creation and delivers deep insights into enterprise account priorities. This dual approach not only expands your pipeline but also reveals who the key stakeholders are within market-leading organizations. Moreover, it strengthens the alignment between marketing and sales, empowering sales teams with tailored messaging for high-value accounts. Investing in developing need-based pipelines for enterprise accounts yields superior returns compared to traditional sales and marketing initiatives.

For Management

Gaining visibility and control over your pipeline enables you to architect growth, rather than simply react to what surfaces. This strategic oversight allows you to prioritize high-impact deals, six-figure investments, and focus resources where they will drive the greatest business value. By anticipating emerging customer needs and market trends, you can proactively shape the organization's growth trajectory and consistently achieve ambitious goals.

For Sales Enablement

Modern sales enablement is about empowering teams to spend more time with customers while ensuring every interaction is informed by actionable insights and aligned with growth strategies. By streamlining workflows and reducing administrative burdens, you enable account teams to focus on what matters most—building relationships and closing deals. A data-driven, organized approach to tracking activities and outcomes transforms sales enablement into a powerful engine for sustained success.

For Commercial Insight Strategists

CISs exist to translate insights into actionable growth opportunities and inform account sellers regarding each customer need, the situation and context, the stakeholders involved, the potential solution, and value message alignment and changes as they occur. This is now an essential role in the workflow of insight-driven account growth. Commercial Insight Strategists will proactively guide account sellers regarding priority opportunities and inform and recommend actions sellers can take throughout an opportunity life cycle.

You've seen how aligning every team around customer relevance and actionable insights can transform your pipeline and set your organization apart. But what does it really take to break away from "the way it's always been done" and win in today's unpredictable market?

Let's step into the shoes of a game changer who rewrote the rules and discover how the same bold, data-driven mindset can revolutionize your approach to B2B growth. Ready to see what happens when you stop following tradition and start building your own playbook for success? Turn the page—your *Moneyball* moment starts now.

CHAPTER 2:

EVERY INSIGHT NEEDS A STORY

've been here before: standing in the dugout, scanning a roster built on star power rather than a systematic approach to scoring runs and racking up wins. The old guard wants big names and flashy stats, but I see an opportunity to rewrite the rules. I dive into the numbers, searching for patterns others miss. What if winning isn't about charisma or the perfect swing, but about getting on base one overlooked player at a time? With each unconventional pick, the tension grows: tradition versus transformation. But as the season unfolds, those forgotten players deliver, and suddenly the impossible feels possible. This is leading with vision, betting on insight over instinct, building a legacy where no one expected one.

The old way, scouting for stars with the "right look," no longer works. That's the world Billy Beane faced in *Moneyball*. Instead of following tradition, he dug into the data, found what really drove wins (getting on base), and built his team around undervalued players. The result? Success against all odds.

Now swap the baseball diamond for the B2B world. For years, companies relied on "star" sales reps to deliver product-centric pitches, believing features and benefits would close deals. But the game has changed. Today's B2B buyers expect sellers to understand their unique needs, not just push a product.

Here's the *Moneyball* lesson for B2B:

- **Move Beyond Gut Feeling:** Like Beane, challenge the "we've always done it this way" mindset.
- **Embrace Data-Driven Insights:** Use data to uncover what truly drives buying decisions, identify high-potential accounts, and understand customer needs.
- **Change the Rules of Engagement:** Focus on solving customer problems and delivering value, not just on pushing products. Analyze buying signals and tailor every interaction.
- **Build a Repeatable, Scalable System:** Beane's process wasn't a one-off; it was a system. B2B leaders must create processes that make success predictable, not dependent on a few "superstars."

The winners today are those who challenge convention, embrace change, and put the customer—not the product—at the center. Like Beane's A's, they use data to see what others miss, adapt faster, and achieve more with less. In the new B2B world, it's not about selling harder; it's about selling smarter and always playing to win.

STORYTELLING & THE HERO'S JOURNEY

Every customer initiative is a journey. In B2B, each initiative is its own journey for the executive buying team, where the path forward is anything but clear. Obstacles loom: risk, internal politics, conflicting priorities, and limited budgets. Progress stalls, momentum fades, and the finish line seems to drift farther away.

But just as a visionary coach sees potential where others see problems, you can help executive buyers find their way. When you illuminate the path to their desired outcomes—when you connect with them in a way that sparks excitement and belief— you become more than a vendor. You become part of their story, a catalyst for transformation.

Gallup's research has found that:

> About 70 percent of decisions are based on emotional factors and only 30 percent are based on rational factors.[4]

Emotionally compelling stories strongly influence the buyers' perspectives, their desired outcomes, and sales.

THE HERO'S JOURNEY BLUEPRINT

Every great turnaround follows a familiar arc. Joseph Campbell called it the hero's journey, a universal story pattern that mirrors the real-world challenges businesses face as they strive for growth and change.

- **Call to Adventure:** Something disrupts the status quo. A challenge emerges, demanding action.

---

4 Ryan Pendell, "Customer Brand Preference and Decisions: Gallup's 70/30 Principle," *Gallup*, accessed September 21, 2025, https://www.gallup.com/workplace/398954/customer-brand-preference-decisions-gallup-principle.aspx.

- **Crossing the Threshold:** The team steps into the unknown, facing new risks and discomfort.
- **Trials, Villains and Allies:** Obstacles arise—resistance to change, silos, competing agendas. Allies and villains appear.
- **The Ordeal:** The toughest tests force the team to adapt, learn, and persevere.
- **The Return:** Success means bringing others along, proving the value of the new world, and embedding lasting change.

> *In today's enterprise landscape, the "hero" leads a buying group, with an average of eleven stakeholders per team, each with their own fears and ambitions.*[5]

This group often comprises senior executives who can override decisions at any time. Navigating this gauntlet requires empathy, insight, and the ability to tell a story that unites and inspires.

STORYTELLING: THE SECRET WEAPON

Just as *Moneyball*'s Billy Beane rewrote the rules by focusing on what truly mattered, you can help your customers see beyond the noise. Storytelling is your playbook. Facts and figures fade, but a compelling narrative endures. When you frame the customer's journey as a hero's quest, full of challenges, villains, allies, and ultimate triumph, you give meaning to the struggle and clarity to the path ahead.

---

5 "Growth Risks 2024: B2B Buying Behaviors Are Evolving." *SBI*, accessed September 21, 2025. https://sbigrowth.com/insights/growth-risks-2024-b2b-buying-behaviors-are-evolving.

- **Guide, Don't Just Sell:** Be the mentor who's seen the journey before, offering wisdom and encouragement.
- **Illuminate the Stakes:** Help buyers see what's possible and what's at risk if they stand still.
- **Celebrate the Transformation:** Make the customer the hero, and their success is the story others want to emulate.

In the end, every successful initiative is a story of vision over convention, of data-driven insight over winging it, and of shared triumph over individual effort. The journey is never easy, but with the right guide, the impossible becomes possible, and a new legacy is born.

TURNING OPPORTUNITY ANALYSIS INTO A COMPELLING NARRATIVE

Breaking down an opportunity is not just a strategic exercise; it is the essential groundwork for crafting a powerful and persuasive story. This analytical process forms the bedrock upon which every successful initiative is built. By dissecting an opportunity, you equip yourself with the clarity and structure needed to engage every stakeholder, align your team, and, ultimately, guide your "hero" toward triumph.

Why Storytelling Matters in Opportunity Analysis

Consider this: Every great story follows a journey, and every successful project mirrors this narrative arc. When you methodically analyze an opportunity, you are, in effect, mapping out the hero's journey for your customer. This approach doesn't just make your message more memorable; it transforms abstract goals into a relatable and motivational adventure for your customer stakeholders.

The Framework: From Analysis to Action

Let's break down how each element of opportunity analysis directly relates to the classic hero's journey, ensuring your initiative resonates on both a rational and emotional level:

Breaking Down an Opportunity	Hero's Journey Parallel
Branded Name for Initiative	The Big Idea—the vision of a better future
The Challenge	The Ordinary World—the present-day reality
The Situation	The Villain—the forces causing disruption
The Hero, Guide, and Wallet Owner	The Protagonists—the key players and VIPs
The Stakeholders	The Ordeals—the conflicts of others involved
Desired Outcomes	The Reward—the promise of a transformed world
The Solution and Value Statements	The Elixir—the solution that brings change
Execution Team	The Allies—those who help realize the vision
Insights	The Wisdom—the lessons that justify the quest

2.1

Use Case—From Dugout to Boardroom: The *Moneyball* Hero's Journey in B2B

Let's step out of the dugout and into the boardroom. The tension between tradition and transformation isn't just a baseball story. It's the universal challenge of every leader facing a changing game. The *Moneyball* scenario is more than a sports tale; it's a blueprint for modern B2B account growth and expansion success, mapped perfectly onto the hero's journey.

1. **The Challenge: The Ordinary World.** This is the current state, and something is challenging it, rendering it no longer desirable. This signal threatens the status quo.

2. **The Situation: The Villain.** The customer is facing a villain that poses a threat to the current state. This threat could be economic, legislative, competitive, environmental, technological, or otherwise.

3. **The Hero, Guide, and Wallet Owner.** In *Moneyball*, Billy Beane is the hero, willing to challenge the status quo, question tradition, and risk his reputation for a better way. In B2B, the hero is the customer leader, who must succeed by changing the status quo and moving toward a better future state. There is an internal guide at the customer's organization to the supplier account executive, and an account executive from the supplier organization that serves as a guide to the hero. Finally, the wallet owner must create a budget for the hero to proceed.

4. **The Stakeholders: Your Customer's Buying Network.** In B2B, there are various operating stakeholders who execute the joint solution. Like the players on Billy Beane's *Moneyball* team, these are the individuals involved in executing the future state.

5. **The Desired Outcomes: The Reward.** What each stakeholder needs to achieve from their perspective. Often, stakeholders' desired outcomes focus more on personal interests than on business results. Therefore, it is important to consider what each individual is truly seeking to accomplish.

6. **The Solution and Value Statements: Create the Path.** Embrace insights to identify what is truly causing the

problem. What is the job that the customer needs to have done? Identify the current state and the pain associated with it. Define the gap and the path from the current state to the future state. Articulate the outcomes to be achieved and the value of transitioning from the current state to the future state. The greater the threat posed by the current state, the greater the impact of the value statements.

7. **The Execution Team: Your Team as Allies to Customer Stakeholders.** The execution team is not just the account seller; it comprises the entire team that aligns with the customer's stakeholders to meet each of their desired outcomes.

8. **The Insights: What *Moneyball* Teaches B2B.** Challenge convention. Customers want new insights they do not yet have about addressing the challenge at hand.

Keep the insights coming. Be the guide to your hero in their journey to success throughout the process.

Continuously adapt the solution and value statements to changing conditions related to the challenge.

LEAD THE CHANGE: HARNESSING THE HERO'S JOURNEY TO INSPIRE, ALIGN, AND ACCELERATE BUSINESS TRANSFORMATION

In business transformation, the hero's journey is not just a story-telling device; it's a proven framework for guiding organizations through change and inspiring action. By methodically break-ing down each stage of this journey, you create a clear "from what to what" narrative that demystifies complex transitions. Yet, as every leader knows, the path is rarely smooth. Obstacles such as resistance to change and entrenched departmental silos are inevitable. This is precisely when the power of storytelling becomes essential. Articulating a compelling vision of the future and communicating its value to every stakeholder can galvanize teams, break down barriers, and foster alignment.

Once you've defined the essential components of your story, it's time to craft your storylines. Think of a storyline as the backbone of your narrative: the sequence of events that brings your vision to life. In B2B environments, the buyer's journey can span six to eighteen months, mirroring the lengthy process of solution implementation and outcome realization. The heart of your story should focus on how you and your customer can jointly overcome a threat or problem and achieve a desired future state and meaningful outcomes.

To develop persuasive storylines, consider these critical elements:

- Where have you encountered and conquered similar challenges before?
- How were these challenges addressed in the past?
- What are the risks of ignoring the issue?
- How does the envisioned future state differ from today's reality?
- What are the key steps and roles involved in the transformation?
- What is the realistic timeline for change?
- Are there distinct phases along the journey?
- What obstacles might arise, and how can they be overcome?
- What outcomes can be expected, and are they sustainable?
- Who were the heroes in past successes, and what made them effective?

For example, imagine a competitor who streamlined redundant processes, freeing account teams from repetitive data entry across multiple platforms. By adopting a new online system, they reclaimed an average of ten hours per week. But the real breakthrough wasn't just in time saved, it was in the renewed collaboration among teams and the noticeable shift in customer engagement. This momentum not only improved pipeline quality but also empowered leadership to showcase new growth indicators, aligning departments and accelerating access to budgets.

Adopting a storytelling mindset enables you to weave these storylines throughout the entire journey, reinforcing your message at every stage. Leveraging AI to structure these narratives, drawing from case studies, proposals, and project plans, can further enhance their relevance and impact. Remember, a compelling storyline is not a product pitch or brochure. It forges both logical and emotional connections to a future state that is energizing and motivating for your audience. When done well, your story becomes a tool that stakeholders use to advocate for change within their own circles, multiplying its influence.

In today's complex business landscape, the ability to tell a powerful, purpose-driven story is not just advantageous, it's essential. Embrace the hero's journey, anticipate obstacles, and inspire your teams and customers to move forward together. The story you tell today can become the catalyst for tomorrow's success, so let's review the core story characters and the elements you need to understand.

THE HERO

Every successful transformation story has a central figure—a hero—who leads the charge from the present state to a promising future. This hero is not just a participant; they are the catalyst, the change agent who inspires, mobilizes, and sustains momentum within their organization. Their journey is challenging, their role demanding, and their impact, when harnessed correctly, is profound.

The Hero's Essential Qualities

To fulfill this pivotal role, a hero must possess more than just a title or authority. They need influence, vision, and, most importantly, an unwavering passion for the initiative. The path to change is rarely smooth. Obstacles will arise, resistance will surface, and setbacks are inevitable. The true hero demonstrates resilience, navigating these ordeals with determination and grace, never losing sight of the goal.

But here's an important truth: **You are not the hero in your customer's story.** Instead, you are the trusted guide, the adviser who helps them navigate the complexities of their journey. Your expertise lights the path, but it is the hero who must walk it.

Lessons Learned from Decades of Change

Having partnered with countless heroes across industries and decades, I've seen firsthand that no two change journeys are alike. Each organization faces its own unique set of obstacles, and each hero brings a distinct set of strengths to the table. Recognizing these differences is crucial to crafting a successful strategy.

Let's break down the common obstacles to change and the attributes a hero must embody to overcome them:

Change Agent Characteristics

Obstacle to Change	What's Needed	Change Agent Characteristics
Complexity	Strategy and Execution	Known for Getting Things Done
Inertia	Motivation	Believes in a Cause
Constraints	Prioritization	Disciplined
Silos	Alignment	Positive Reputation
Disorder	Structure	Detail-Oriented
Fragmentation	Sponsorship	Natural Organizer
Confusion	Engagement	Patient Educator
Volatility	Resilience	Withstands Difficult Situations

2.2

Often these challenges appear in combination, requiring the hero to draw upon a blend of these characteristics. It's essential to assess your hero's strengths and the specific obstacles they face. As a guide, you can help identify gaps and recommend ways to fill them, either from within the customer's team or by leveraging your own resources. What change agent characteristics does your customer hero embody? Take some time to consider their strengths and gaps.

Recognizing and Empowering the Hero

Spotting a true hero is both an art and a science. When you find someone with the strength, perseverance, and vision to drive change, invest in them. Their leadership is the engine that propels the opportunity forward.

But how do you identify a great hero? Here's a simple, actionable checklist:

Change Agent (Hero) Attributes

Influencer Attributes	Score 1 - 3
Trusted internally by executive team	
Politically connected across organization	
Can obtain budget	
Can obtain authority	
Previously led change initiatives	
Has a strategic role regardless of title	
Is tenured or has been hired to lead change	
Successful track record & promotion history	
Believes in causes that are good for company	
Has executive level authority	
Is respected by peers	
Leadership / coaching characteristics	
Has vision & execution	
Motivated by personal achievement	
Total	

1 = Lowest, 3 = Highest

2.3: Hero Checklist

Score your hero against these criteria and revisit this assessment as the journey unfolds. By doing so, you maximize your chances of success and ensure you're supporting the right champion. If your hero is not strong enough and you can't effectively guide them, the initiative will stall.

Empowering Your Internal Champion

When a deal stalls, it is often because your internal champion, the "hero," has lost momentum. As a strategic partner, it is essential to assess whether you are supporting someone with the influence and drive to see the initiative through to completion. Remember, your involvement is only one part of a larger effort. For your customer, selecting your solution is not the final objective. Their true goal is to achieve measurable business outcomes with minimal risk and maximum return, in collaboration with you.

It is crucial to recognize that your champion's professional reputation and sometimes their career trajectory may hinge on the success of this project. The most effective account managers are those who identify the right champion and invest their energy in high-impact activities that empower them to succeed within their organization. By aligning your efforts with your champion's internal objectives, you not only accelerate the deal but also build a foundation for lasting partnership and mutual success.

THE GUIDE

Just as every great transformation hinges on a heroic figure willing to embrace change, so too does every success story require a trusted guide, an expert who lights the path ahead. The guide's influence is subtle yet indispensable, providing not only direction but the wisdom and confidence to avoid pitfalls and press forward when challenges arise. Far more than a bystander, the guide shapes outcomes through empathy, insight, and a steady presence. When harnessed with care, the guide's impact extends well beyond simple advice, enabling others to rise above uncertainty and unlock their fullest potential.

Lead with Insight: The Power of the Guide

In every transformative journey, the presence of a guide is not just helpful, it is essential. Without guidance, even the most capable individuals risk becoming overwhelmed, making costly mistakes, and progressing more slowly than necessary. A skilled guide offers more than advice; they provide expertise, perspective, and tools that empower others to overcome obstacles and accelerate their growth.

The Guide's Role in Business Success

In the world of business and account sales, the guide's influence is often as critical, if not more so, than the hero's efforts. As a guide, you are not merely a mentor; you must be the driver for your customer's transformation. Your responsibility extends beyond offering direction: You must help your customers recognize the realities of their situation, align with their internal teams, and navigate organizational complexities.

Often this means identifying and mentoring an internal advocate within your customer's organization. By equipping this internal guide with your knowledge and experience, you empower them to drive momentum, make informed decisions, and achieve the best possible outcomes. In other words, you can empower another internal guide in addition to yourself.

Curiosity and Insight: Your Tools for Impact

To be effective, approach each engagement with the curiosity of a detective seeking to uncover every relevant detail. Ask probing questions to understand the internal dynamics and "inside baseball" that influence your customer's journey. Think of successful transformations depicted in stories like *Moneyball*, where the guide's insights were pivotal in overcoming inertia and achieving breakthrough results.

Remember, in complex B2B environments, buying journeys are long and fraught with uncertainty. A steady stream of insights from a trusted guide is vital for keeping decision-makers focused, informed, and confident in their actions.

Essential Qualities of an Effective Guide

To maximize your impact as a guide, cultivate the following attributes:

- **Deep Knowledge:** Possess a thorough understanding of the challenges your customers face and the pathways to overcome them.
- **Storytelling:** Use stories to convey wisdom and experience, making complex ideas relatable and memorable.
- **Contextual Awareness:** Help customers see their journey within a broader context, giving their efforts greater meaning and direction.
- **Engagement and Motivation:** Build emotional connections through storytelling, inspiring perseverance and resilience.
- **Relatability:** Share experiences that allow customers to see themselves in the narrative, fostering trust and a sense of shared purpose.

Embrace the Opportunity

Today's rapidly evolving business landscape demands guides who are collaborative, insightful, and proactive. By stepping confidently into the role of the guide, you not only help customers navigate their challenges, but you also become the driving force behind their success. Embrace this responsibility with curiosity, empathy, and a commitment to continuous learning, and you will inspire others to achieve outcomes they never thought possible.

How to Align as the Guide to the Hero

As emphasized in chapter 1, adopting the right mindset is foundational when positioning yourself as the guide to your hero. Remember, your role is not to overshadow the hero, but to empower them throughout their journey. This journey is often long and filled with unexpected challenges, so it's essential to consider diverse and creative ways to support the hero as they navigate their path.

Guiding the Hero

An effective guide understands and
connects with the hero, the change agent.

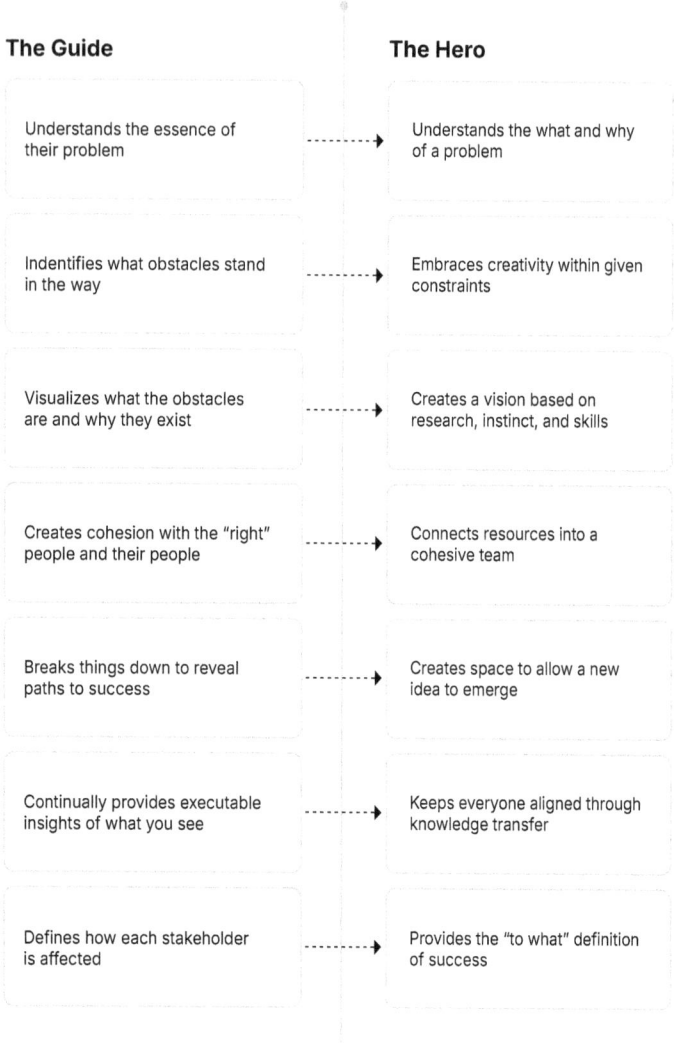

The Guide   **The Hero**

Understands the essence of their problem	Understands the what and why of a problem
Indentifies what obstacles stand in the way	Embraces creativity within given constraints
Visualizes what the obstacles are and why they exist	Creates a vision based on research, instinct, and skills
Creates cohesion with the "right" people and their people	Connects resources into a cohesive team
Breaks things down to reveal paths to success	Creates space to allow a new idea to emerge
Continually provides executable insights of what you see	Keeps everyone aligned through knowledge transfer
Defines how each stakeholder is affected	Provides the "to what" definition of success

2.4: Guiding the Hero

THE WALLET OWNER

Every successful initiative relies on more than just visionary leadership or determined execution. It also hinges on the engagement of those who hold the power to fund, approve, or scale the initiative. Enter the wallet owner: the key decision-maker whose support can open doors or bring progress to a halt. While their involvement may sometimes seem distant, their influence is felt at every major juncture. Ensuring the wallet owner is engaged, informed, and aligned is not an afterthought; it's a strategic imperative. This guide explores practical ways to actively involve the wallet owner throughout the journey, transforming them from a passive gatekeeper into a collaborative champion of your initiative.

Engage the Wallet Owner Throughout the Journey

Never underestimate the importance of the wallet owner. If they are unaware of your role and contributions during the initiative, you risk facing an uphill battle for support at the end. Proactively engage the wallet owner by doing the following:

- **Sharing Actionable Insights**: Offer information that is immediately useful to them, not just status updates.
- **Reporting on Progress**: Regularly communicate key milestones, especially those tied to return on investment and risk reduction.
- **Enabling Advocacy**: Equip your hero or guide with relevant updates and insights to pass along, ensuring your value is recognized at every level.

Consider the *Moneyball* analogy: Billy Beane, the hero, needed to keep the team owner informed about his strategy and execution. The guide, Peter Brand, provided insights that empowered Billy to have those critical conversations. Similarly, your success depends on making the wallet owner an informed ally, not a last-minute audience.

THE VILLAIN

In every great buyer journey involving change, the story is not defined by just the hero but by the obstacles they must overcome. At the heart of every meaningful change lies a formidable adversary—a villain—that fuels urgency and sharpens the focus on what needs to be done. This villain is more than just a roadblock; it embodies the disruptive force that threatens the current state and creates pain, demands action, and compels organizations to move from complacency to action.

Identify and Label the Villain

By understanding and clearly defining the villain in your narrative, you lay the groundwork for collective purpose and inspire the commitment necessary to achieve a lasting impact. The villain may appear differently to each stakeholder, but it is always the same disruptive force. By clearly labeling the villain, you do the following:

- **Simplify Your Narrative:** Focus your story on overcoming what's intolerable.
- **Align Stakeholders:** Reinforce who the enemy is and why defeating it is essential, fostering unity and urgency.

In today's business landscape, inefficiency is often the villain. With AI's growing capability to eliminate inefficiencies, organizations are shifting from "growth at all costs" to sustainable, profitable growth. Inefficiency exemplifies a modern B2B villain, one that must be clearly identified and described from each stakeholder's perspective, so everyone recognizes the challenge.

Make the Villain Real and Relevant

Think of the villain as a negative health condition in a business. Just as people visit doctors to prevent or address health issues, organizations launch initiatives to avoid or respond to business threats. The villain is the trigger—a condition that, if resolved, leads to a desired outcome such as less pain, improved performance, or reduced risk.

- **Diagnose Before Prescribing:** Like a doctor, first identify and define the problem before offering solutions.
- **Tailor the Story:** Play with different villain archetypes to see which resonates most with your stakeholders. Validate your choice to ensure it is both the most common and the most feared enemy.

Motivate Action

Be deliberate and strategic in engaging the wallet owner and labeling the villain. By doing so, you not only clarify the path forward but also inspire collective action. Remember, the clarity and strength of your story can be the difference between passive interest and active support. Make the wallet owner your advocate and ensure everyone knows exactly what you're fighting against and why it matters.

STORY-DRIVEN SUCCESS: BRANDING, VISION, AND THE POWER OF PERSUASION

Every transformative initiative begins not only with a heroic leader but also with a clear and compelling story that captivates and aligns everyone involved. Beyond inspiring the hero who drives change, it is essential to craft a narrative that distinguishes your initiative in the marketplace and deeply resonates with your customers' needs and objectives. This story forms the foundation upon which trust is built, decisions are guided, and momentum is sustained.

To achieve this, branding your initiative with precision, articulating the current and future state, personalizing the vision of success, leveraging relatable analogies, and anticipating obstacles become indispensable steps. Together, these elements create a strategic framework that elevates you from a vendor to a trusted guide, empowering your customer to confidently lead their organization toward a promising future.

Brand the Initiative: Stand Out and Align

Branding your initiative is not just valuable, it is essential. A branded initiative becomes the narrative your customer buys into, shaping their journey with you. Ask yourself: Are you positioned as a core partner in this journey? Do your company's credentials and expertise directly align with the initiative's objectives? Establishing this alignment elevates your team above the competition and cements your relevance.

To truly connect, speak your customer's language. Understand and adopt their terminology. Immerse your team in their internal communications. By mirroring their language and mindset, you build trust and position yourself as an indispensable partner.

Define Current and Future State: Paint a Compelling Contrast

One of the most persuasive communication strategies is to juxtapose the current state with the envisioned future state clearly. This side-by-side comparison highlights gaps, opportunities, and the tangible benefits of change and the path to getting there. Use visuals and concise narratives to help your customer vividly see the transformation ahead.

Can you articulate the current and future state simply and compellingly? Your goal is to inspire excitement and buy-in, especially from your key decision-makers. Remember the pivotal scene from *Moneyball*, where Peter Brand reframes the game for Billy Beane: "Owners should not be interested in buying players; they should be interested in buying wins." This moment distills a complex transformation into a memorable, motivating vision. Strive to create a similarly powerful story for your customer.

Create a Vision for Success: Make Outcomes Personal

Building on your future state, craft a vision of success that resonates with each stakeholder. Move beyond generic value statements. Instead, tell a story that makes the outcome tangible and memorable.

For example, rather than stating, "You'll achieve a 25 percent reduction in risk," share a relatable success story: "A peer in your industry faced the same challenge. By mapping their process, we uncovered an 80/20 gap, a small inefficiency causing the majority of their problems. Together, we closed that gap and delivered a 25 percent risk reduction. We can do the same for you." Stories like these stick. They motivate action and foster belief in the journey ahead.

Use Relatable Analogies: Make It Stick

Analogies are powerful tools for making complex ideas accessible and memorable. Draw from popular culture or industry examples, like the *Moneyball* story, to connect with your audience. Don't hesitate to leverage AI tools to find fresh, relevant analogies—test them with your internal champions to ensure they resonate.

Anticipate and Address Obstacles: Be the Trusted Guide

Anticipating and proactively addressing the internal obstacles your customer, the hero, will face is not just valuable; it's essential for guiding them successfully through their buying journey. Imagine the impact when, instead of waiting for funding challenges to derail progress, you equip your customer in advance with a compelling funding asset that clearly demonstrates the superior value of your solution compared to alternatives. This proactive approach signals that you are more than a vendor; you are a trusted guide, invested in their success, and ready to help them navigate internal hurdles with confidence.

It's crucial to recognize that the materials buyers truly need often differ dramatically from the standard product-focused materials most sellers provide. While sellers tend to offer product overviews, competitive comparisons, and case studies, executive buyers are searching for requirements documents, conceptual solution frameworks, and technical schemas. They want clear road maps, risk analyses, key performance indicators (KPIs), and evaluation guides—tools that speak directly to their internal decision-making processes and business objectives, not just your product's features. See chapter 10 for more details on what customers really want in pitches.

This distinction becomes even more pronounced when

cross-selling or upselling within existing accounts. Here buyers' needs are tightly aligned with their internal priorities and challenges, not generic product literature. To succeed, you must tailor your assets to each initiative, making them not only relevant but also highly repeatable across similar opportunities. Remember, delivering these assets isn't solely the responsibility of the account manager; it's a team effort involving everyone the customer trusts and interacts with throughout the relationship.

The lesson is clear: If you want to empower your customer, the hero, to overcome internal obstacles and drive successful outcomes, you must step into the role of a strategic guide. Analyze their journey, anticipate their roadblocks, and arm them with the right assets at the right time. By doing so, you don't just facilitate a transaction; you build trust, deepen relationships, and position yourself as an indispensable partner in their ongoing success.

LEAD THE JOURNEY: INSPIRING CUSTOMER SUCCESS THROUGH STORYTELLING AND TEAM ALIGNMENT

Every customer success story is driven by an account leader who not only manages the process but inspires the journey itself, someone who weaves the many perspectives of their team and client into a compelling narrative of progress and transformation. This leader understands that storytelling is more than just recapping facts; it is about igniting genuine enthusiasm, fostering alignment, and guiding all stakeholders through the inevitable challenges toward a shared vision. By embracing their role as both a storyteller and a strategist, they transform customer relationships from transactional interactions into meaningful partnerships built on trust, collaboration, and shared success.

What Inspires You in Telling the Story

True storytelling excellence begins with your own genuine enthusiasm. Ask yourself: What energizes you about this journey with
your customer? In every account team, a diversity of roles and
perspectives converge, both internally and with the client. Identify your unique vantage point. What makes the story compelling to you, and why will it resonate with your audience? When
you approach each interaction with authenticity and a sincere
commitment to solving your customer's challenges, you become
not just a service provider but a trusted adviser. This proactive
engagement often leads to deeper relationships and greater loyalty,
as customers recognize you have their best interests at heart.

Shaping the Story for All Customer Stakeholders

Once you've crafted your initial narrative, the real work begins:
adapting and evolving the story as the journey unfolds. Busi-
ness-to-business relationships are rarely linear; unexpected
obstacles will arise and the path will shift. Regularly align with
your team to assess progress and recalibrate. The hero's journey
provides a powerful framework, reminding us that every initiative
cycles through familiar phases: challenge, adversity, and transformation. Your role is to guide, communicate insights, and equip
your customer (the hero) with the tools they need to bring others
along. Map out the key storylines for each stakeholder, ensuring
they all connect to the overarching narrative. Maintain a shared
space where the team can track themes and progress, fostering
alignment and clarity.

What Is the Story for Your Team?

Never forget: Your team is part of the story. They are the guides, walking alongside the stakeholders they match up with on this journey. Teams that see themselves as united guides rather than isolated contributors unlock greater impact. Internal storytelling is a powerful leadership tool; when leaders articulate a clear, shared vision, teams internalize their roles in driving customer success. Imagine the journey as a treasure map: The destination is valuable, but unpredictable obstacles will test your resolve. Use analogies and narratives to help your team visualize their progress and reinforce alignment with the customer's future state. Stories inspire action and cohesion; ensure your team stays motivated and moves forward together toward the shared goal.

ROLE BY ROLE: STEPS TO SHAPE THE FUTURE

For Account Teams

Understanding your customer's journey and positioning yourself as a trusted guide along that path is both strategic and invaluable. Every engagement is not a single transaction, but a chapter in a larger story you share with your customer. Ask yourself: How many journeys will you take together? When you connect through cohesive storylines rather than just data points, you foster deeper engagement and sustained momentum. This approach not only differentiates you but also builds lasting partnerships that extend well beyond the initial sale. Embrace storytelling as your most effective tool for guiding stakeholders toward shared success.

For Marketing

Storylines are the heartbeat of impactful marketing. By crafting compelling narratives that illustrate the urgent need for change, you elevate your firm's value in the eyes of your customers. Sharing these storyboards internally ensures that every team member communicates a unified, strategic vision. This not only clarifies your role in helping customers achieve their desired outcomes but also positions your organization as an essential partner in their journey. Remember, marketing that tells a story inspires action and fosters loyalty.

For Management

Executive communication is most effective when it reinforces the overarching team narrative. When management engages with their counterparts, whether at the chief executive officer (CEO) level or beyond, it's essential to convey storylines that align with the broader organizational vision. Invest time in understanding both the desired end state and your current position in the buyer's journey. Empathy and sincerity are nonnegotiable, especially in C-suite interactions, as they build the trust necessary for strategic alignment and long-term success. Lead by example and champion the power of authentic storytelling.

For Sales Enablement

Developing storytelling capabilities is not just a skill; it's a strategic imperative. Encourage account teams to weave narratives into their daily interactions, making each conversation more persuasive and memorable. Adopt a structured approach: Tie storylines directly to customer initiatives and active opportunities, document successful stories, and facilitate role-playing to refine these narratives for different stakeholders. Leverage AI tools to enhance and scale your storytelling efforts. By embedding storytelling into your sales culture, you empower teams to influence decisions and drive growth.

For Commercial Insight Strategists

The opportunity reports you create for account executives follow the hero's journey to facilitate storytelling. When you develop value statements and a solution narrative for each opportunity, look for the story inside. The better you can pull out storylines that resonate with account executives, the more effective they will be in articulating these with enthusiasm.

You've just discovered how storytelling can transform your influence, elevate your impact, and forge deeper connections with your customers. But here's the thing: Every great story faces unexpected twists, and every hero encounters challenges that test their resolve.

As you step into the next chapter, imagine yourself not just as a guide, but as a navigator through storms of change—market shifts, economic upheavals, and technological revolutions that can appear without warning. The real adventure begins when you learn how to help your customers thrive, not just in calm waters, but when the winds of uncertainty are at their fiercest.

So, are you ready to meet the heroes and allies who will shape your journey and to discover how to lead them through the unpredictable business climate ahead? Let's dive into the next chapter.

CHAPTER 3:

UNDERSTANDING THE CUSTOMER'S BUSINESS ENVIRONMENT

In every **B2B buying journey,** there are four essential characters: the hero (our customer), the guide (you as the account leader, and you have a guide within the customer organization), the wallet owner, and the broader team. Over my career, I've learned that while these individuals drive the process, they are constantly buffeted by external forces, much like athletes trying to win a football game as tornadoes sweep across the field. The business environment today is more volatile and unpredictable than ever before, and these shifting conditions can stall or even derail the most promising deals.

THE UNPREDICTABLE WEATHER OF BUSINESS

Let's reflect on the impact of COVID-19, which began in 2020 and continues to shape our world. When the pandemic first hit, I, like many, assumed it would be a much briefer disruption. Yet, as months dragged on, it became clear that we were facing a profound and lasting transformation. The uncertainty was staggering. Masks, remote work, and the loss of in-person connection became our new normal. The ripple effects on business were immediate and far-reaching.

Just as the team in *Moneyball* had to reinvent itself to stay competitive, businesses during the pandemic had to rapidly adapt, finding new ways to drive revenue and meet evolving customer needs. For many, this meant doubling down on key accounts, prioritizing cross-selling and upselling as strategic levers for growth.

ADAPTING TO SURVIVE—AND THRIVE

But adaptation didn't stop there. The demand for commercial efficiency has become paramount. The era of "growth at all costs" has given way to a focus on profitable, sustainable growth. The emergence of AI tools like ChatGPT in 2022 is a testament to this shift. Forward-thinking organizations are leveraging artificial intelligence to streamline operations, boost productivity, and uncover new opportunities for efficiency.

And let's be honest, COVID-19 was just one storm among many. Inflation, immigration, supply chain disruptions, cybersecurity threats, global conflicts, shifting demographics, and constant legislative changes all add layers of complexity to the business landscape. Each of these factors can dramatically affect resource allocation, customer priorities, and, ultimately, the success of your deals.

MY ADVICE: PREPARE, ADAPT, AND LEAD

If there's one lesson I've learned, it's this: Resilience and agility are now essential qualities for every account team and seller. You must anticipate change, stay close to your customers, and be ready to pivot your strategy as conditions evolve. Don't wait for the storm to pass; learn to play and win in the rain.

Embrace innovation. Invest in technology. Prioritize your most valuable accounts. And, above all, foster a mindset of continuous learning and adaptation within your team. The storms will keep coming, but with the right approach, not only can you weather them; you can emerge stronger and more successful than ever before.

NAVIGATING BUSINESS STORMS: GUIDING YOUR HERO THROUGH VUCA CONDITIONS

Every business, regardless of size or industry, is subject to a wide array of external forces. These environmental conditions, ranging from technological shifts to legislative changes, shape the landscape in which companies operate. Consider the following influences:

- Industry trends
- Competitive dynamics
- Artificial intelligence
- Digital transformation
- Technological advancements
- Legislative and regulatory changes
- Financial market fluctuations
- Evolving customer expectations
- Political instability

- Geographical factors
- Climate and environmental risks
- Supply chain disruptions

Each of these factors acts like a weather system, sometimes calm, sometimes stormy, and always capable of affecting your customer's business and its stakeholders. When these conditions shift, they don't affect just the organization; they also impact your "hero" (the key buyer or decision-maker), every stakeholder involved in the purchasing process, and your own account team.

The Hero's Journey in a Storm

Imagine your hero embarking on a complex buying journey. This path is rarely straightforward. Instead, it is fraught with sudden storms: VUCA. These forces can disrupt schedules, alter priorities, shift stakeholder involvement, and change the very scope and timing of decisions.

To truly support your hero, you must do more than simply acknowledge these challenges. You must understand the specific "storm conditions" they face and adapt your approach accordingly. This is where your role as a guide becomes vital.

The Origin and Power of VUCA

The concept of VUCA—volatility, uncertainty, complexity, and ambiguity—originated with the US Army in the late 1980s.[6]

---

6 "Where Does the Term VUCA Come From," *VUCA World*, accessed September 21, 2025, https://www.vuca-world.org/roles-of-nanus-and-bennis/.

Faced with unpredictable and rapidly changing global situations, military leaders needed a framework to assess and respond to the chaos of modern warfare. VUCA provided a way to identify the nature of disruption and to adapt strategies in real time, improving the likelihood of mission success.

VUCA has been embraced by the business world as a powerful lens for understanding and navigating the unpredictable forces that shape markets and organizations.

Applying VUCA to Business: Your Role as the Guide

Just as military leaders use VUCA to assess and respond to battlefield conditions, business leaders and account teams can use this framework to map out the challenges facing their heroes. By analyzing the type and intensity of disruption, be it technological upheaval, political instability, or supply chain breakdowns, you can provide a "weather map" for your hero's journey.

Why This Matters

- **Insightful Guidance:** Understand the specific VUCA conditions at play so you can offer relevant, timely insights that empower your hero to make informed decisions.
- **Adaptive Storytelling:** Tailor your narrative to address the unique challenges your hero faces, increasing your credibility and value as a trusted adviser.
- **Proactive Support:** Anticipate obstacles and help your hero and their stakeholders navigate through uncertainty with confidence and clarity.

By equipping yourself with insightful guidance, adaptive storytelling, and proactive support, you position yourself as a true navigator in the face of volatility and uncertainty. But what truly sets exceptional leaders and advisers apart in today's unpredictable environment?

THE POWER OF INSIGHT: NAVIGATING COMPLEXITY IN MODERN BUSINESS

Imagine standing at the helm of your organization as the business landscape shifts beneath your feet. Markets fluctuate, competitors pivot, and new technologies emerge at breakneck speed. In this environment, what separates those who merely survive from those who thrive? The answer is insight.

What Is an Insight?

At its core, an insight is the ability to see beyond the obvious—to pierce the fog of uncertainty and grasp the deeper truths that drive people, markets, and organizations. It is the power of understanding a situation so thoroughly that new opportunities become visible, and decisive action becomes possible.

But not all insights are created equal. Actionable insights go a step further: They illuminate a clear path forward, revealing opportunities to change, adapt, and achieve your most ambitious goals.

Companies that leverage an insights-driven sales approach are more likely to achieve significant growth. Forrester says:

> *Insight-driven companies are 8.5 times more likely to create at least 20 percent year-over-year growth.[7]*

---

7 Srividya Sridharan and Gene Leganza, "Build an Insights-Driven Business: The Executive Overview of the Insights-Driven Business Playbook," *Forrester*, January 27, 2022, https://www.forrester.com/report/build-an-insights-driven-business/RES139876.

Types of Insights

Patterns
Recurring themes or behaviors that reveal underlying dynamics.

Trends
Emerging directions that signal where the market or industry is headed.

Discoveries
New findings that challenge assumptions or open new avenues.

Correlations
Connections between variables that inform strategic choices.

Anomalies
Outliers and exceptions that may signal risk or opportunity.

Events
Significant occurrences that reshape the competitive landscape.

Results
Outcomes that validate or challenge your strategies.

3.1: Types of Business Insights

Business insights come in many forms, each offering a unique lens through which to view your challenges:

- **Patterns:** Recurring themes or behaviors that reveal underlying dynamics.
- **Trends:** Emerging directions that signal where the market or industry is headed.
- **Discoveries:** New findings that challenge assumptions or open new avenues.
- **Correlations:** Connections between variables that inform strategic choices.
- **Anomalies:** Outliers and exceptions that may signal risk or opportunity.
- **Events:** Significant occurrences that reshape the competitive landscape.
- **Results:** Outcomes that validate or challenge your strategies.

Increasingly, humans are leveraging AI to generate insights with unprecedented speed and context, transforming raw data into strategic advantage.

Leading with Insight and Agility

To thrive in this environment, account teams must do more than react—they must anticipate. Your role as a guide is to help your hero navigate through the storms by understanding the unique VUCA conditions they face and adapting your strategy accordingly. This means offering insightful guidance, tailoring your approach to evolving challenges, and providing proactive support that empowers decision-makers.

Note: Please refer to Appendix 1 to see a sample list of VUCA driven business conditions and insights. This is provided as a representative sample of what businesses often face.

By leveraging technology, fostering a culture of continuous learning, and prioritizing your most valuable relationships, you position your team to not just weather the storms—but emerge stronger and more successful. Embrace change, invest in innovation, and lead with confidence. The storms will keep coming, but with the right mindset and approach, you can guide your hero to victory in even the most turbulent conditions.

ROLE BY ROLE: STEPS TO SHAPE THE FUTURE

For Account Teams

Buying executives are highly receptive to new, actionable insights, yet frequently lack the time to discover these opportunities themselves; by proactively sharing the valuable perspectives you uncover, presented in concise summaries with clear alerts highlighting their significance and pointing out what merits attention, you not only strengthen internal alignment but also establish yourself as a trusted adviser to your customer stakeholders. This approach sets you apart as a dedicated resource who prioritizes their best interests, encourages timely responses, and often inspires tangible, forward-looking action. Adopting this advisory role demands consistent awareness, thoughtful communication, and a genuine commitment to adding value, but when done with clarity, strategic intent, and a solution-oriented mindset, it not only elevates your professional credibility but also drives meaningful results for all parties involved.

For Marketing

To elevate your organization's impact, insights drawn from customer accounts must be not only gathered but meticulously analyzed for trends and recurring themes that truly resonate with key decision-makers. Marketing plays a pivotal role in this process by translating these insights into compelling messages and content that shape your company's narrative and strengthen its position in the marketplace. By consistently seeking out and leveraging account-specific information, marketing can ensure that communications are not only broadly relevant but also deeply meaningful to both existing customers and prospective clients. Adopting this strategic approach encourages alignment with buying executives' priorities, fostering greater engagement, and reinforcing your credibility as a thought leader, so stay proactive, attentive, and ready to turn every insight into an opportunity that drives business growth.

For Management

Running your business relies on insights, and what's important to your customers should be important to you. Look at what customers consider important when making decisions about where you should monitor industry and account events. The biggest companies in an industry serve as a bellwether for the rest of the market and provide a good indication of where customers are heading.

For Sales Enablement

Keep an eye on major areas of interest in key accounts, as well as the types of problem statements and threats these represent. This information can help inform account teams about new priority areas of focus. Sharing these insights across teams is valuable because progress depends not on product sales but on addressing customer needs and challenges as they arise, and the ability to act on them.

For Commercial Insight Strategists

You are the first to see new insights emerge and can identify when these should be packaged and sent to customers. This process is highly valuable. Creating external, customer-facing insight reports, as well as internal reports to align your solutions, value statements, and storylines, is another important service you can provide to account executives.

Before we dive in, let me ask you: What if you could do more than just survive the storms? What if you could actually harness their energy to propel your business forward? The truth is, every challenge you face is also an opportunity to deepen your customer relationships and shift from simply reacting to truly leading.

So how do you move from weathering the chaos to orchestrating real, lasting impact with your accounts? The answer lies in discovering and mastering the unique rhythms within each relationship. Ready to transform everyday transactions into powerful partnerships? Let's turn the page and unlock the strategies that will set you apart.

CHAPTER 4:

SEEING AROUND CORNERS INSIDE YOUR ACCOUNT

If you aren't growing the account,
you are losing the account.

This isn't just a catchy phrase. It's a hard-earned truth I've witnessed firsthand throughout my career. Early on, I learned that complacency is the silent adversary of account management. The reality is that decisions that shape the future of your accounts are often made by stakeholders who rarely interact with your team daily.

Let me share what keeps me vigilant: Countless conversations and decisions happen behind closed doors—sometimes with your competitors, sometimes about changes in strategy, budget cuts, or even removing your primary contacts. The unsettling part? These shifts are rarely telegraphed in advance.

- Are stakeholders quietly considering a switch?
- Is your budget on the chopping block?
- Are your champions being replaced or sidelined?
- Do decision-makers truly value your partnership, or are they just going through the motions?

These are the questions that keep account leaders up at night. Unless you have deep insight into the customer's mindset, you're operating in the dark.

STAY AHEAD OR FALL BEHIND: WHY PROACTIVE CUSTOMER INTELLIGENCE IS YOUR ONLY DEFENSE

Unexpected shifts in customer behavior aren't just theoretical risks. I've confronted them firsthand. The sinking feeling that comes with losing a key account or hearing about a major change too late has taught me that complacency is costly. While it's tempting to believe that consistent communication or healthy reports mean all is well, this surface-level information rarely tells the full story. The reality is, major setbacks often result from issues that were quietly brewing undetected, beneath the surface of what the account team can see. That's why I've learned to approach customer relationships with a sense of urgency and deeper curiosity, always looking to uncover what might otherwise remain hidden.

The Cost of Being Caught Off Guard

I've experienced the shock of an unexpected customer decision. It's not just unsettling; it's a scramble for answers that often don't exist, because the warning signs were never visible. When it comes to your most valuable customers, ignorance isn't just risky, it's potentially catastrophic. Losing a major account can erase months, even years, of sales gains in a single blow.

Why Traditional Metrics Fall Short

Many companies rely on metrics like net promoter score (NPS) to gauge customer health. But for complex B2B relationships, this is a superficial measure at best. It fails to capture the nuanced agendas, shifting priorities, and evolving objectives of enterprise customers with hundreds of stakeholders. The advisory firm Gartner predicts that:

> *Over 75 percent of organizations will abandon NPS as a customer success measure by 2025.*[8]

A Call to Action: Proactive Relationship Management

Here's my advice. Don't wait for a crisis to realize you're too out of touch. Invest in robust monitoring and engagement processes that go beyond surface-level metrics. Build relationships at every level of the customer organization. Seek regular, candid feedback. Strive to understand not just what your customers are saying, but what they're planning. And do this consistently over time.

The most successful account leaders are those who treat growth as a mandate, not an option. Stay curious, stay connected, and never assume the status quo is secure. Your vigilance today is the best insurance against tomorrow's surprises.

---

8 Harvard Business Review Analytic Services, "Beyond Net Promoter Score: Customer Experience Measurement Reimagined," research report (sponsored by Genesys), 2022, https://hbr.org/resources/pdfs/comm/Genesys/BeyondNetPromoterScore.pdf.

WHAT ARE CUSTOMERS REALLY THINKING?

Let's be honest: Customers think about themselves, both as individuals and as organizations. Their perspective is fundamentally different from that of a vendor. While vendors focus on their products, services, and messaging, customers are laser-focused on their own goals, challenges, and daily realities. Every decision is shaped by their unique objectives, evolving business conditions, internal changes, shifting priorities, and external pressures.

In the previous chapter, we explored the broader forces shaping the customer environment. Now let's take a closer look at the specific stakeholders—by role and department—and examine how they perceive your actions, your relationship, and your value.

The Customer's Lens: It's Not About You

Here's a critical insight: customers value partners who demonstrate a genuine interest in helping them succeed. Unfortunately, this runs counter to how many organizations train their sales teams. Too often, sellers are sent into the field, armed with product-centric messages that do little to build trust or credibility. The result?

According to Gartner:

> *75 percent of business executives would prefer to avoid interacting with salespeople altogether.[9]*

This is not just a statistic—it's a wake-up call. When the very people responsible for driving revenue are being systematically shut out, it's clear that something needs to change.

---

9 Gartner, "The B2B Buying Journey: Key Stages and How to Optimize Them," *Gartner,* accessed October 2, 2025, https://www.gartner.com/en/sales/insights/b2b-buying-journey.

A Cautionary Tale: Losing Sight of Stakeholders

Let me share a story that underscores the risks of neglecting stakeholder perceptions. My company once lost a major client—not because of poor performance or inadequate service, but because we failed to engage with key decision-makers at higher levels. Our day-to-day contacts were satisfied, but decisions were being made elsewhere, influenced by competitors who had seized the opportunity to build relationships where we had not.

By the time we realized what was happening, it was too late. The client's transition to a competitor created chaos, not only for us but for the client and the competitor as well. Ironically, the new solution was inferior, and the disruption was entirely avoidable. The lesson was clear: Failing to monitor and shape the perceptions of all stakeholders at every level puts even your strongest relationships at risk.

WHY YOUR BIGGEST ACCOUNTS DEMAND A TAILORED STRATEGY

Every organization has a select group of accounts—the top ten, twenty, or fifty—that serve as the true engines of revenue and growth. This isn't just a coincidence; it's what I call the "economy of key accounts." Consider this:

> *The world's largest companies, the Global 2000, generate revenue that exceeds 50 percent of the world's gross domestic product (GDP).*[10]

---

10 Forbes Corporate Communications, "Forbes Announces 22nd Annual Global 2000: A Ranking of the World's Largest Companies," *Forbes* (press release), June 13, 2024, https://www.forbes.com/sites/forbes-spotlights/2024/06/13/forbes-announces-22nd-annual-global-2000-a-ranking-of-the-worlds-largest-companies/.

Their sheer scale, influence, and spending power create a vast divide between them and smaller businesses. Over time, these enterprise giants become only more expansive and complex, often representing the largest and most critical customers for any firm. Just look at the meteoric rise of the Magnificent 7 tech companies—each year, their impact and reach set new benchmarks for what's possible in business. According to McKinsey & Company:

> *The Magnificent Seven added $247 billion to global economic profit from 2020-2024.*[11]

That is global economic profit, not top line revenue.

Given this reality, it's essential to treat your biggest accounts differently. No company can afford to lose one of its largest customers. I've seen this firsthand: At a previous company, ten customers accounted for 70 percent of revenue and 50 percent of profit out of a portfolio of more than twelve hundred customers. Meanwhile, the bottom three hundred customers cost the company money. We dedicated specialized teams to our top accounts and kept them in constant focus. When it came time to "make the quarter," these were the accounts we turned to—every time.

But here's where the story gets even more compelling: The potential for further growth within these accounts and among their peers is enormous. A consulting firm once analyzed our business and revealed that by focusing on just 135 companies and forty-five hundred key relationships across five to seven buying centers, we could triple our revenue. It sounded simple. In practice, it was anything but. Why? Because true enterprise

---

11 Marc de Jong and Peter Stumpner, "Global Economic Profit Bounces Back to an All-Time High," *McKinsey & Company*, September 4, 2025, https://www.mckinsey.com/capabilities/strategy-and-corporate-finance/our-insights/global-economic-profit-bounces-back-to-an-all-time-high.

growth and relationship development require more than CRM systems and product training. They demand deep, actionable insights—understanding how executives think, how decisions are made, and how your company is perceived.

According to Gartner:

> *Customers who feel confident in their decision-making are 2.6 times more likely to buy more from their existing vendors.[12]*

Yet there's no substitute for the account team's ability to capture and act on this intelligence. This need for insight is a double-edged sword. On one side, it's about managing risk—ensuring your largest accounts don't walk away. On the other, it's about unlocking new growth opportunities. Let's first explore the risks: What are the warning signs that your top accounts might be at risk of leaving, and how can you proactively address them?

The stakes are high, but so are the rewards. By treating your biggest accounts with the strategic focus and tailored approach they deserve, you not only safeguard your core revenue but also lay the groundwork for profitable growth. Now is the time to elevate your key account strategy because, in today's economy, your future depends on it.

UNDERSTANDING AND MANAGING RISK IN KEY ACCOUNT RELATIONSHIPS

When it comes to managing key accounts, risk is an ever-present companion. Whether you're a seasoned account manager or new

---

12 Gartner and Kelly Blum (Contributor), "What Sales Should Know About Modern B2B Buyers," *Smarter With Gartner*, November 5, 2019, https://www.gartner.com/smarterwithgartner/what-sales-should-know-about-modern-b2b-buyers.

to the field, recognizing and addressing these risks is not just prudent—it's essential for long-term success. Let's explore the landscape of key account risk, understand its sources, and discover practical ways to turn potential pitfalls into opportunities for deeper partnership.

The Story Behind Key Account Risk

Imagine you're steering a ship through unpredictable waters. Your customers are the valuable cargo, and your organization is the vessel entrusted with their journey. The risks you face are like hidden reefs. Some are within your control; others are shaped by the currents of your client's business. Navigating these waters requires vigilance, adaptability, and a proactive mindset.

To chart a safe course, we must first understand where these risks originate. Broadly, they fall into two categories: **client-driven factors** and **supplier-driven factors.** Let's break these down and explore how you can address each with confidence and foresight.

Client-Driven Risk Factors

1. **Concentration of Spend**
   - **Risk:** The more a client invests with you, the more attention—and scrutiny—you attract. Clients may worry about putting too many eggs in one basket.
   - **Action:** Proactively diversify your engagement across different departments and functions. This not only spreads risk but also deepens your footprint within the organization.
2. **Complex Stakeholder Environments**
   - **Risk:** Multiple departments mean multiple perspectives, priorities, and decision-makers.

- **Action:** Align your team members with stakeholders who share similar backgrounds and expertise. This builds rapport and ensures your message resonates at every level.

3. **Senior-Level Involvement**
    - **Risk:** The higher the stakes, the higher up the decision chain you must go, often facing new challenges and unfamiliar faces.
    - **Action:** Prepare your executives to engage in meaningful conversations with their counterparts. Senior-level buy-in is crucial for securing large, strategic deals.

4. **Competitive Pressures**
    - **Risk:** When competitors are present, the risk of losing ground increases—sometimes due to perceptions rather than reality.
    - **Action:** Regularly share performance data and competitive benchmarks. Transparency builds trust and positions you as the partner of choice.

5. **Leadership Turnover**
    - **Risk:** New leaders bring new preferences and loyalties, potentially threatening established relationships.
    - **Action:** Act swiftly to build relationships with incoming executives. Shape their perception of your value from day one.

Supplier-Driven Risk Factors

1. **Inconsistent Service or Support**
    - **Risk:** Service lapses can quickly erode trust, and negative experiences tend to overshadow positive ones.
    - **Action:** Overcommunicate during challenging times.

Show genuine commitment to resolving issues and improving the customer experience.

2. **Failure to Meet Performance Expectations**

   - **Risk:** Clients continuously evaluate whether you deliver on promises. Perceived underperformance can lead to lost business.

   - **Action:** Conduct regular performance reviews, highlighting measurable impacts. If gaps exist, address them proactively and visibly.

3. **Operational Complexity**

   - **Risk:** Cumbersome processes can frustrate clients and make you difficult to work with.

   - **Action:** Empathize with customer frustrations and seek internal solutions. Aggregate feedback to advocate for systemic improvements.

4. **Misalignment with Customer Needs**

   - **Risk:** Assuming you know what the client wants can be a costly mistake.

   - **Action:** Listen deeply and ask probing questions. Treat each client's situation as unique and avoid jumping to conclusions.

5. **Lack of Proactive Recommendations**

   - **Risk:** If you're seen as reactive rather than consultative, competitors may outshine you.

   - **Action:** Make innovation a regular agenda item. Monthly or quarterly reviews should focus on your client's evolving needs, not just your own offerings.

6. **Fragmented Customer Experience**

   - **Risk:** Disjointed communication leads clients to view you as a mere vendor, not a strategic partner.

- **Action:** Foster a collaborative, well-informed team. Share insights and coordinate efforts to deliver a seamless, unified experience.

Turning Risk into Opportunity

The difference between knowing what to do and doing it can be vast. But here's the good news: Every risk factor is also an opportunity to demonstrate your value, commitment, and adaptability.

Be proactive. Be empathetic. Be relentless in your pursuit of partnership excellence.

By understanding the sources of risk and addressing them head-on, you not only protect your key accounts but also lay the groundwork for deeper, more resilient relationships. Remember, the most successful account teams aren't those who avoid risk, but those who navigate it with insight, agility, and a genuine desire to help their clients succeed.

UNLOCKING GROWTH: THE REAL DRIVERS OF OPPORTUNITY WITH KEY ACCOUNTS

What truly creates growth opportunities within key accounts? The answer is both straightforward and surprisingly overlooked: Most opportunities originate with the customer. Yet many teams remain unaware of this reality, focusing inwardly rather than tuning into the evolving needs and challenges of their clients.

Imagine an "invisible pipeline" running through every major account (as discussed in chapter 1)—a constant flow of unmet needs, emerging problems, and untapped potential. The teams that learn to see and navigate this pipeline are the ones who achieve expansive, sustainable growth. I've witnessed organizations draft hundreds of contract add-ons over a few years, and they excel by consistently focusing on understanding and addressing customer needs.

Let's dive into the key areas where opportunities arise, analyze their causes, and explore how you can proactively capture them.

1. **Expanding Reach Across Departments and Business Lines**
   - **Growth Factor:** The more access you have across departments and lines of business, the greater your growth potential. When you combine broad reach with strong references and existing contracts, you're poised for significant expansion.
   - **Opportunity Capture:** Map out the key executives across all buying centers. Understand their relationships, dependencies, and shared influencers. Systematically leverage your existing connections to forge new ones and make follow-up a disciplined habit. Too often, account teams become comfortable with a handful of relationships and miss the bigger picture. Don't fall into this trap! In one case, a sales engagement team cultivated more than 125 relationships, while the account team itself had fewer than five, highlighting the risk of a narrow focus. Broad, strategic relationship building is essential.
2. **Navigating External Disruption**
   - **Growth Factor:** Industries and customers are constantly disrupted by external forces—regulatory changes, technological shifts, and economic cycles. Each disruption creates new needs and problems to solve.

- **Opportunity Capture:** Stay vigilant. Track these events, categorize them, and map out where to focus your growth efforts. Customers respond to disruption by launching funded initiatives. If you're organized, proactive, and relevant, you'll secure a larger share of the available budget.

3. **Capitalizing on People Changes**
   - **Growth Factor:** Every personnel change—whether a promotion, departure, or reorganization—creates new dynamics and opportunities. New leaders bring fresh priorities and budgets.
   - **Opportunity Capture:** Monitor these changes closely. React quickly to understand what each shift means for the affected stakeholders. New responsibilities often translate into new spending—and new opportunities for you to add value.

4. **Leveraging Subsidiary Networks**
   - **Growth Factor:** Large enterprise accounts often own hundreds of subsidiaries, each with their own decision-makers and budgets.
   - **Opportunity Capture:** Identify the key stakeholders and buying centers within the most relevant subsidiaries. Prioritize your outreach and work systematically through the list. There is often a vast, untapped pipeline here that goes unnoticed.

5. **Tapping into Global Presence**
   - **Growth Factor:** Global organizations operate in multiple countries, each with its own leadership, budgets, and priorities.
   - **Opportunity Capture:** Map out the global leadership

structure, at least in the top countries relevant to your business. Build relationships with these leaders to unlock new avenues for growth.

6. **Responding to Mergers & Acquisitions**
   - **Growth Factor:** Mergers and acquisitions reshape priorities, create new needs, and open areas for consolidation and growth.
   - **Opportunity Capture:** Develop a clear perspective on the value of these combinations. Engage key stakeholders early to understand how the changes affect them and to identify new gaps and opportunities.

The Path Forward: Vigilance, Engagement, and Value

Growth and risk factors are always present within key accounts. Your job is to monitor, understand, and act on them to protect and expand your position. Ask yourself:

- Do you truly understand the strategies and plans of your account executives across the organization?
- Are you building trust with all the stakeholders who matter?
- Are you shaping their perception of your company in a positive, consistent way?
- Are you demonstrating reliability and value at every touchpoint?
- Would your most important contacts recommend you to their peers?
- Are your competitors building relationships with your key contacts while you stand still?

The story of key account growth is written by those who stay curious, act boldly, and never take a relationship for granted. The opportunities are there if you're willing to look beyond the obvious and commit to continuous, strategic engagement.

IMPLEMENTING AN IMMERSIVE & ADAPTIVE GROWTH STRATEGY: A ROAD MAP FOR ACCOUNT SUCCESS

Imagine you're navigating a winding road, full of unexpected threats and opportunities just out of sight. To truly "see around corners" in your client relationships, you must move beyond the status quo and embrace an immersive, adaptive approach, one that keeps you proactive, agile, and deeply connected to your customers' evolving needs.

Let's explore the three pillars of this strategy, each designed to help you not only sustain but accelerate account growth, while building trust and differentiation in today's competitive landscape.

1. Ask with Purpose: The Power of Insightful Inquiry

The most successful relationships are built on understanding. Too often, organizations assume they know what their stakeholders want, only to be blindsided by shifting priorities or silent dissatisfaction. The antidote? Ask, and keep asking.

Think of these questions as your regular "health check" for the relationship:

- Are we meeting your expectations? Please elaborate.
- Are we delivering the value you anticipated? Your feedback is invaluable.
- Is it easy to do business with us? We'd love your thoughts.
- Are we responsive to your needs? Please share examples.

- How would you describe the impact of our partnership on your business?

Go further by probing into the future:

- What upcoming priorities or challenges can we help you address?
- What new goals should we be aware of?
- How would you rate our communication—both frequency and clarity?

By inviting open-ended feedback, you gain a treasure trove of actionable insights. Stakeholders appreciate being heard, especially when the focus is on their needs rather than your offerings. This approach not only differentiates you from competitors who rarely ask but also signals your genuine investment in their success. Remember, a simple Net Promoter Score won't capture the nuances of a long-term B2B relationship; you need a more robust, ongoing dialogue.

2. Engage with Empathy: Building Authentic Connections

Business relationships thrive on trust, and trust is earned through empathy and authenticity. The best account executives understand this intuitively, but it's essential to cultivate this mindset across your entire team.

Ask yourself and your colleagues:

- Are we truly empathetic to each stakeholder's unique needs?

- Do we listen actively and follow through on our commitments?
- Are we curious about the root causes of their challenges?
- Do we approach problem-solving from their perspective, not just our own?

These are not innate traits. They're learned behaviors, honed through consistent practice and a willingness to learn from every interaction. Beware of the temptation to dominate conversations with product pitches or to overlook follow-up; both can erode trust and credibility. As one experienced leader remarked, "We spend 90 percent of our time talking about our product and only 10 percent understanding the problem. We need to flip that." Make understanding the foundation of your engagement.

3. Model and Map: Navigating the Customer's World

To align with your clients' business objectives, you must become adept at modeling their enterprise, understanding the intricate web of stakeholders, decision-makers, and influencers that shape every buying decision.

Consider these critical questions:

- How do we categorize and analyze relationship health scores?
- Are we tracking changes and adapting our approach accordingly?
- Can we synthesize feedback to uncover emerging patterns and themes?
- Are we quick to address issues ("put out fires") and identify new opportunities ("fertile soil")?

- How do we support their decision-making process and bolster their confidence?

Spotting patterns is not just about data; it's about storytelling. When you organize business challenges by buying center and correlate them with stated goals, you begin to anticipate needs and proactively support your clients' growth. This level of insight allows you to navigate complex networks with confidence and agility.

Reflect on these areas:

- What are the individual and collective goals of the buying group?
- Are there common threats across different stakeholders?
- What decision-making patterns are emerging?
- Does everyone involved feel understood and represented?
- How confident are they in their decisions, and how can you help reinforce that confidence?

The Journey Ahead

Adopting an immersive and adaptive growth strategy is not just a tactical move; it's a mindset shift. It requires curiosity, discipline, and a relentless commitment to understanding and serving your clients. The rewards are substantial: deeper trust, greater loyalty, and a sustainable path to mutual growth.

As you embark on this journey, remember: The organizations that thrive are those that listen deeply, engage authentically, and adapt fearlessly. Be the partner who sees around corners—and leads the way forward.

UNLOCKING THE POWER OF RELATIONSHIP RHYTHMS

Have you ever wondered why some account relationships flourish while others seem to stall or drift? The answer often lies in how we recognize and manage the natural patterns within these relationships, what we call relationship rhythms.

A relationship rhythm is a powerful framework for understanding the dynamic nature of account management. Just as music relies on rhythm to create structure and flow, successful account relationships depend on identifying and nurturing their own unique patterns. Relationships within enterprise accounts are especially complex, often evolving, shifting, or even deteriorating as circumstances change. By organizing these relationships into distinct types, you gain clarity and control, making it far easier to monitor, track, and influence outcomes.

Through extensive analysis, we've identified four primary relationship rhythms: Responsive, Accommodative, Prescriptive, and Performance. Each rhythm represents a different stage and style of engagement, from the transactional to the strategic. Think of Responsive and Accommodative Rhythms as product-driven and more transactional, while Prescriptive and Performance Rhythms are outcome-driven and deeply strategic.

Why does this matter? Because when you recognize the rhythm of a relationship, you can tailor your approach—anticipating needs, addressing challenges proactively, and driving meaningful results. This structured perspective not only streamlines your efforts but also empowers you to build stronger, more resilient partnerships.

As you navigate your portfolio of accounts, challenge yourself to identify the rhythm at play. Are you merely reacting, or are you setting the tempo? By mastering relationship rhythms, you don't just manage accounts; you orchestrate lasting success.

Relationship Rhythm Types

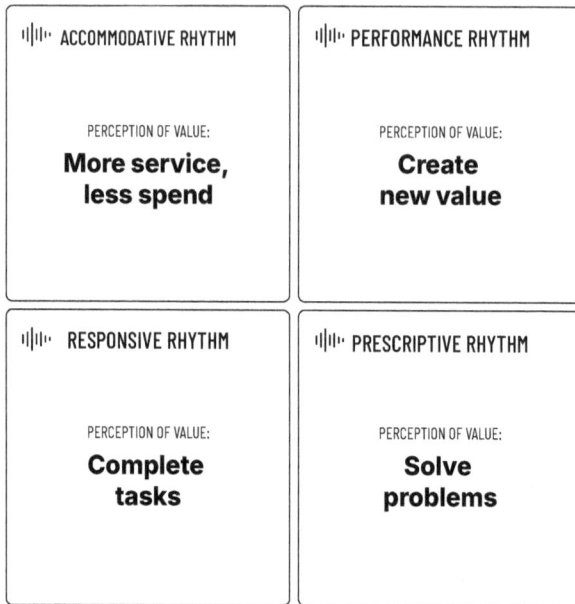

ılılı ACCOMMODATIVE RHYTHM	ılılı PERFORMANCE RHYTHM
PERCEPTION OF VALUE: **More service, less spend**	PERCEPTION OF VALUE: **Create new value**
ılılı RESPONSIVE RHYTHM	ılılı PRESCRIPTIVE RHYTHM
PERCEPTION OF VALUE: **Complete tasks**	PERCEPTION OF VALUE: **Solve problems**

4.1: Relationship Rhythm Chart

FROM VENDOR TO VISIONARY: MASTERING THE FOUR RELATIONSHIP RHYTHMS TO TRANSFORM CUSTOMER ACCOUNTS INTO STRATEGIC PARTNERSHIPS

Every organization that delivers products or services operates within a set of distinct relationship rhythms, representing a stage in the journey from transactional supplier to strategic partner. Understanding these rhythms is not just an academic exercise; it's essential for any business leader determined to drive sustainable growth, deepen customer loyalty, and unlock new value from key accounts.

Responsive Rhythm: Mastering Speed and Reliability

At the core of every successful business relationship is the Responsive Rhythm, a dynamic, foundational stage where speed and precision in meeting customer demand take center stage. This is often the first impression a company makes, shaping the customer's perception of value through the swift and reliable fulfillment of orders and resolution of service needs. Imagine the seamless process of ordering from Amazon: Efficiency and responsiveness are not just appreciated; they are expected.

In this rhythm, automation and inside sales teams become the backbone of operational excellence. They streamline interactions, from sales to fulfillment and support, ensuring that every transaction is processed with minimal friction. As artificial intelligence and digital tools raise the bar for what is possible, customers come to expect nothing less than flawless, high-volume service at every touchpoint.

However, while excelling in the Responsive Rhythm is essential, it comes with a caution: Businesses that linger here risk becoming indistinguishable from their competitors, valued only for their speed and price rather than for deeper, more strategic contributions. To avoid being seen as a mere commodity provider—easily replaced by a faster or cheaper alternative—organizations must recognize when it's time to evolve. Mastering the Responsive Rhythm is a powerful starting point, but true differentiation requires moving beyond efficiency to build lasting, value-driven partnerships.

RESPONSIVE RHYTHM

MEASURED BY: **High Volume / Reduced Labor Costs**

DESCRIPTION:	OPPORTUNITY TYPE:	CUSTOMER EXPECTATIONS:
Order Taking	**Repeat Purchase**	**Speed & Reliability**

SCALED BY: **Automation**

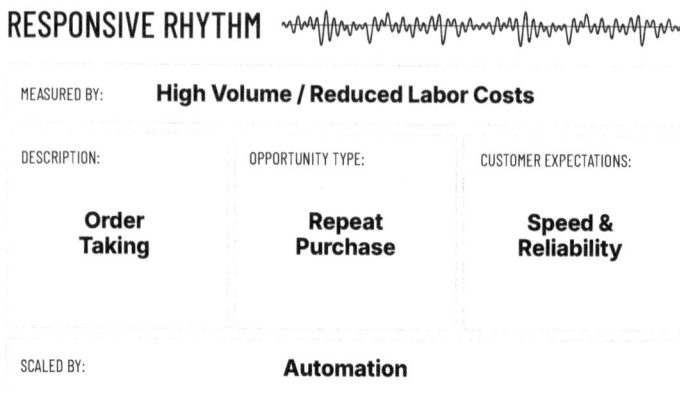

4.2: Responsive Rhythm Chart

The Hidden Risk: Procurement's Relentless Pursuit

The significant risk lurking beneath the surface involves procurement departments that are constantly on the lookout for cost savings, and relationships governed by the Responsive Rhythm are often their primary targets. Their mission is clear: Demonstrate value by lowering prices. This means your supplier relationship is under continual scrutiny, with procurement teams ready to renegotiate or replace you at the first sign of inefficiency or complacency.

The Growth Imperative: Service Excellence as a Differentiator

How do you safeguard and grow these relationships? The answer is deceptively simple: Deliver exceptional service. While some may argue that great service is no longer enough, experience shows that it remains a powerful differentiator, especially when relationships are at risk. It's crucial to ensure that your support team not only meets but exceeds customer expectations. If a team member isn't the right fit, make a change swiftly. The stakes are high; even the largest contracts can unravel if customers lose confidence in your people or your service.

The Critical Success Factor: Relentless Customer Feedback

One of the most overlooked strategies is the continual pursuit of customer feedback. Make it a habit to ask your stakeholders, "Is it easy to do business with us?" and, just as importantly, "What could we do better?" Then act on their recommendations. Embedding this feedback loop into your daily operations is essential. Too many organizations become complacent, assuming that strong relationships will endure without ongoing effort. Don't fall into this trap.

The Path Forward

To truly excel in the Responsive Rhythm, you must combine speed and reliability with a relentless commitment to service and continuous improvement. Challenge yourself and your team: Are you making it easy for customers to choose you, time and again? Are you proactively seeking ways to raise the bar? By embracing these principles, you not only protect your position, you also create opportunities for growth and lasting partnership.

Accommodative Rhythm: Mastering Integration and Value Creation in Evolving Business Partnerships

As business relationships evolve, they often transition into what we call the Accommodative Rhythm—a pivotal stage where the supplier's role expands beyond simple transactions to aggregating demand and broadening the scope of products and services offered. In this rhythm, customers' expectations shift: They seek more comprehensive solutions, greater efficiency, and often anticipate achieving these benefits while reducing their overall spend. This rhythm is not merely about selling more; it demands a sophisticated understanding of the customer's evolving needs and a proactive approach to adapting your offerings. Imagine a supplier who can seamlessly consolidate purchase volumes, unlocking significant discounts and streamlining procurement—this is the hallmark of the Accommodative Rhythm. When executed with precision, it elevates the supplier to a critical partner, driving value through both scale and service.

However, with this increased integration comes heightened responsibility. As customers place greater trust in their suppliers, the expectations for exceptional service, transparency, and adaptability intensify. Sellers must excel not only in delivering competitive pricing but also in communication, coordination, and responsiveness. Every interaction, from sharing delivery timelines and managing product updates to navigating price changes and optimizing bundles, requires elevated expertise and unwavering commitment. Embracing the Accommodative Rhythm is both an opportunity and a challenge: It is a chance to become indispensable, but only for those willing to meet the rising demands with agility, insight, and professionalism.

ACCOMMODATIVE RHYTHM

MEASURED BY: **Volume Purchase Agreements / Improved Efficiency**

DESCRIPTION:	OPPORTUNITY TYPE:	CUSTOMER EXPECTATIONS:
Bulk Buys	**Volume Discounts**	**Service & Responsiveness**

SCALED BY: **Analytics**

4.3: Accommodative Rhythm Chart

Navigating the Risks: The Threat of Commoditization

The significant risk lurking beneath the surface: commoditization. When products or services become indistinguishable from those of competitors, margins erode, competition intensifies, and customer loyalty wavers. In such an environment, switching costs are low, and it becomes all too easy for customers to seek alternatives. The challenge is clear: How do you stand out and protect your business from being reduced to just another commodity?

A Path to Growth: Delivering a Frictionless Customer Experience

The answer lies in relentlessly pursuing a frictionless customer experience. While some may argue that automation alone is insufficient, real-world evidence shows that businesses who invest in seamless, automated processes build enduring customer relationships. By continuously refining and automating every touchpoint, you not only enhance efficiency but also elevate the overall value you provide. This commitment to excellence becomes your differentiator, earning you the right to deepen and expand your relationship with each customer.

The Critical Success Factor: Proactive Engagement and Advocacy

Success in this environment demands more than operational excellence; it requires a proactive, customer-centric mindset. Regularly ask your stakeholders, "Are we meeting your needs?" and be prepared to adapt quickly to obstacles, problems, and emerging opportunities. Don't just respond—anticipate. Offer cost-saving recommendations before they are requested and ensure that both your customer and their procurement teams know you are actively monitoring and optimizing service and value.

Transform your customers into advocates by consistently measuring and acting on their feedback. When procurement sees that you are as invested in their success as they are, you not only secure your position—you set the stage for the next level of partnership.

By embracing these principles, you can navigate the complexities of volume-based relationships, mitigate the risks of commoditization, and build a foundation for sustainable growth. The journey is demanding, but the rewards—for both you and your customers—are well worth the effort.

Prescriptive Rhythm: Elevating Customer Engagement

In the dynamic landscape of business relationships, the Prescriptive Rhythm represents a shift to a problem-solving, consultative-level relationship. In this rhythm, account teams move beyond simply reacting to client needs; they proactively identify and solve problems, becoming indispensable advisers. Customers begin to rely on your team's insight and reliability, valuing your ability to anticipate challenges, understand their unique business context, and collaborate on innovative solutions. Trust becomes the cornerstone of your partnership, and your recommendations carry significant weight. This is the moment when you move from being a vendor to becoming a true strategic ally.

Picture the seller as a trusted physician. Rather than merely offering remedies, the prescriptive adviser listens intently, diagnoses underlying issues, and prescribes tailored solutions with precision. This is the essence of the Prescriptive Rhythm: a disciplined, consultative approach that transforms sales from a transactional exchange into a genuine partnership. In today's environment, success demands more than product knowledge. It requires deep expertise, empathy, and a commitment to guiding customers through complexity.

The need for such partnerships is urgent. Consider that:

> *85 percent of CEOs acknowledge that even their own executive teams struggle to overcome their organizations' toughest challenges.*[13]

This reality underscores a profound truth: Buyers are not just searching for products or services—they are actively seeking informed partners who can help them navigate uncertainty and drive meaningful change. Embracing the Prescriptive Rhythm positions you to meet this need, forging relationships built on trust, insight, and lasting value.

---

13 Thomas Wedell-Wedellsborg, "Are You Solving the Right Problems?," *Harvard Business Review*, From the Magazine January–February 2017, https://hbr.org/2017/01/are-you-solving-the-right-problems.

PRESCRIPTIVE RHYTHM

MEASURED BY: **Sizeable, Profitable Contracts / Improved Workflows**

DESCRIPTION:	OPPORTUNITY TYPE:	CUSTOMER EXPECTATIONS:
Problem Solving	**Treatment Recommendations**	**Informed Guidance Expertise**

SCALED BY: **Knowledge Management**

4.4: Prescriptive Rhythm Chart

Why Most Account Teams Fall Short

Most account teams still operate at the surface level, missing the opportunity to become indispensable advisers. Customers are hungry for insights and actionable know-how. They want partners who can illuminate blind spots, especially when the stakes are high or the issues are urgent. Those who can deliver this level of consultative engagement become trusted allies, not just vendors.

The Greatest Threat: Informed Competitors

There is a significant risk looming for those who lag behind. If you approach customers uninformed, focused solely on your own agenda, or fail to add genuine value, you will quickly find yourself sidelined. In today's landscape—where AI-powered insights are rapidly raising the bar—being late or irrelevant is not just a disadvantage; it's a recipe for exclusion from the executive table. The harsh reality is that well-informed competitors are ready to step in and seize the opportunity.

The Path Forward: Embrace Insight-Driven Growth

The key to growth is clear: Become relentlessly insight-driven and obsess over your customer's success. This requires a shift in mindset and rhythm—moving from a product-centric approach to a customer-first orientation. Insights are the currency of relevance. When you deliver timely, prescriptive recommendations, you earn trust. Trust, in turn, unlocks higher-value opportunities and long-term, profitable relationships.

Ask yourself: Are you present and relevant in the moments that matter most to your customers? If not, you risk missing out on the very deals that could define your success.

The Heart of Success: Actionable Insights

But what exactly is an insight? It's not just a headline or a fleeting signal of need. True insights are forged by connecting the dots—linking the challenge at hand, the stakeholders involved, the urgency and value of resolution, and the orchestration of resources toward a shared vision of success.

To win in this environment, organizations must empower their account teams with a steady stream of actionable insights. Make it a daily habit to equip your team with the knowledge and context they need to engage customers meaningfully and in real time. This is not a one-time effort, but a continuous discipline that separates the best from the rest.

Performance Rhythm: Elevating Account Relationships

At the apex of account management is the **Performance Rhythm**—a stage where relationships transcend traditional problem-solving and evolve into true business partnerships. Here the focus shifts to **cocreating new value through innovation and**

collaboration, enabling both you and your customer to achieve outcomes that would be impossible independently. This rhythm is not only dynamic and strategic but also mutually transformative, demanding a foundation of deep trust and a willingness to explore uncharted territory together. Leading consulting and professional services firms exemplify this approach by assembling diverse, specialized teams to design tailored solutions, rather than simply offering off-the-shelf products. By continuously generating fresh, measurable ideas, these organizations foster enduring client relationships and drive repeat business at scale. Embracing the Performance Rhythm is both an opportunity and a responsibility: It accelerates growth, sets new standards for collaboration, and positions your organization as an indispensable partner in your client's ongoing success.

The Biggest Risk: Falling into Reactivity

A cautionary note—remaining reactive is the greatest threat to achieving this high-value relationship. If you rely solely on customers to identify their needs, you risk being perceived as a commodity rather than as a strategic partner. Senior executives, accountable for delivering shareholder value, are drawn to well-conceived ideas that demonstrate foresight and impact. To earn their trust, you must lead with insights, proactively presenting innovative concepts that resonate due to their potential scale and significance.

PERFORMANCE RHYTHM

MEASURED BY: **Maintain & Grow Wallet Share / Optimized Resource Allocation**

DESCRIPTION:	OPPORTUNITY TYPE:	CUSTOMER EXPECTATIONS:
Innovative	**Ideation**	**Cross-functional Collaboration**

SCALED BY: **Collaborative Teams**

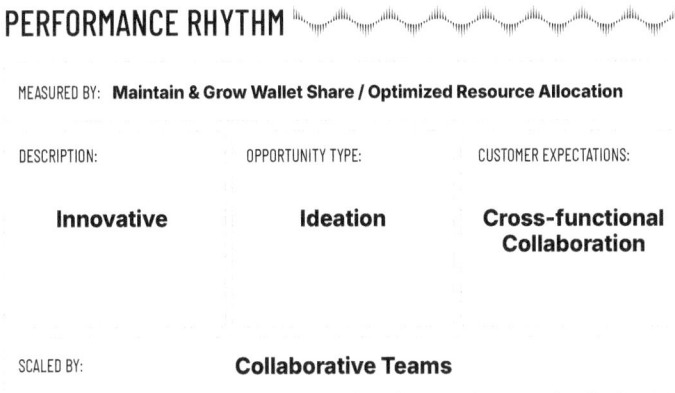

4.5: Performance Rhythm Chart

Growth Strategy: Making Collaboration Your Competitive Edge

True growth is fueled by collaboration. Research from Gartner Group highlights that:

> *Chief sales officers rank the lack of internal collaboration as a top challenge.[14]*

If your teams struggle to work together, it becomes nearly impossible to bring a united front to customers. Leaders must champion a culture of collaboration, breaking down silos and encouraging cross-functional engagement to unlock the full potential of collective expertise.

Critical Success Factor: Centralizing Account Intelligence

Success hinges on the ability to centralize and continuously

---

14 Michael Katz, "3 Top Trends for Chief Sales Officers in 2025," *Gartner*, May 28, 2025, https://www.gartner.com/en/articles/trends-for-chief-sales-officers.

update account intelligence, making it accessible to every team member. Account planning should be a shared responsibility, not confined to the seller alone. Encourage regular collaborative sessions focused on uncovering insights and brainstorming innovative business concepts to present to customers. This collective intelligence not only sharpens your competitive edge but also ensures that every interaction adds value.

Mastering Account Rhythms: Transforming Customer Relationships from Transactions to Strategic Partnerships

It's important to recognize that within any sizable account, these rhythms may coexist. One division may rely on you for rapid, responsive service, while another looks to you for strategic innovation. The imperative is to assess your current position within each relationship and chart a deliberate course toward higher-value and more profitable engagement.

For those committed to growth, the challenge and the opportunity are clear: Don't settle for the status quo. Diagnose where you are, envision where you want to be, and engineer your approach to elevate each relationship to its fullest potential. By mastering these rhythms, you not only meet customer expectations, you exceed them, turning everyday transactions into lasting partnerships that fuel long-term success

MAPPING YOUR ACCOUNT GROWTH STRATEGY

Map where you stand today and a vision for where you want to go. The Growth Strategy Dashboard serves as your compass, helping you chart a course through the complex landscape of client relationships and opportunities.

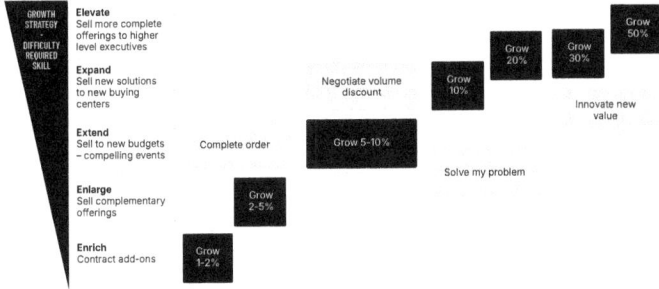

4.6: Growth Strategy Dashboard

Understanding Relationship Rhythms

Consider the different rhythms that define your interactions with an account:

- **Responsive and Accommodative Rhythms** are tactical, focused on immediate needs and product delivery.
- **Prescriptive and Performance Rhythms** are strategic, centered on solving problems and driving innovation.

When you map your account relationships on the dashboard, ask yourself: Where are your current opportunities clustered? It's common to find most engagements concentrated in the lower left—where deals are smaller and margins thinner. The upper right, in contrast, represents larger, more lucrative opportunities that demand a more strategic approach.

The Role of Contractual Agreements

Beyond relationship types, examine the contracts shaping your account. The structure of these agreements influences every-thing—from commercial value and success metrics to resource allocation, duration, and commitment. Each contract is more than just a legal document; it's the blueprint for a commercial relationship that can evolve and strengthen over time.

Overlaying Bluespace Analysis

Leverage the Bluespace analysis we discussed in chapter 1 to identify not only where you are, but where you could go. This overlay helps you visualize both current business and untapped opportunities. To guide your thinking, reflect on these pivotal questions:

- Do you have the relationships necessary to pursue your targeted opportunities?
- What steps will move you from one relationship type to another?
- How should your team be structured to establish credibility with your buyer?
- Which opportunities offer the quickest route to revenue?
- What recurring challenges and needs does the account present?
- Who are the key stakeholders at various organizational levels?
- What will your first steps be to expand these relationships?

Navigating Stakeholder Dynamics

Relationship rhythms aren't limited to the account as a whole. They exist with individual stakeholders too. Accurately labeling the type of relationship you have with each key contact is essential.

Work backward: Define the relationship you aspire to have, then chart the actions needed to get there. As you do, you'll notice that as relationships evolve, so do the executives involved in decision-making.

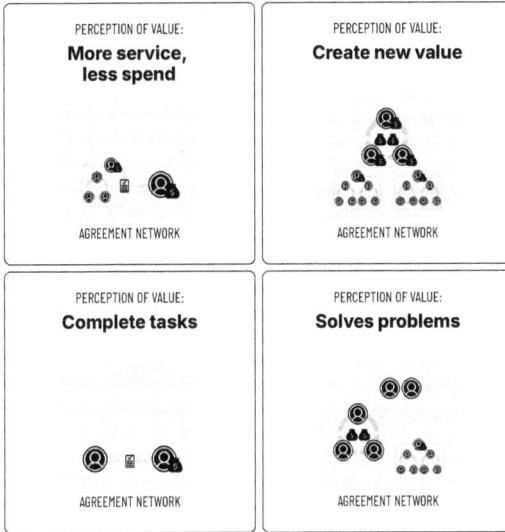

4.7: Navigating Stakeholder Dynamics

Building Strategic Buyer Networks

As you move toward more strategic rhythms, the network of stakeholders grows. These "buyer networks" are responsible for major decisions and initiatives. The stakes rise for both sides, making it crucial to approach these relationships with thoughtful planning and disciplined execution—random, uncoordinated engagement simply won't suffice.

Elevating Your Influence

Having organized your opportunities by rhythm and assessed your current relationships, you're now poised to expand your influence. The next step is to build robust "spheres of influence" around stakeholders, positioning yourself and your team to elevate and deepen these connections.

This is your opportunity to transform risks into opportunities, anticipate challenges, and become a trusted partner in your client's growth. With a clear strategy and a commitment to purposeful engagement, you're ready to unlock the full potential of your account relationships.

ROLE BY ROLE: STEPS TO SHAPE THE FUTURE
For Account Teams

Mastering both risk assessment and the identification of growth opportunities is fundamental to effective enterprise account management. As you gather and catalog stakeholder perspectives, you unlock the ability to map each individual and opportunity to distinct relationship rhythms. Think of these rhythms as your strategic GPS—they guide you in nurturing and advancing your account relationships with intention and precision. This disciplined approach is not just a best practice; it is a cornerstone of long-term success as an account leader. By embedding these rhythms into your daily operations, you equip yourself to execute growth strategies with greater confidence and impact. Remember, deliberate relationship management isn't just about today's wins—it's about building a resilient foundation for tomorrow's achievements.

For Marketing

Deep insights into how stakeholders perceive the health of your account relationships, and their broader interests, are invaluable assets for shaping go-to-market strategies. These perceptions illuminate the real problems, priorities, and sources of value that matter most to your audience. By analyzing relationship perceptions by role and industry, you can craft messaging that truly resonates and develop case studies, success stories, and testimonials that speak directly to your market's needs. Patterns in these perceptions often reveal the underlying causes behind customer decisions, providing you with a powerful lens to anticipate trends and drive more effective marketing initiatives. Harness this intelligence to not only tell compelling stories but create lasting connections with your audience.

For Management

Customer perception is the bedrock of your business's reputation and trajectory. Understanding what your customers need and value from you directly informs research and development, product innovation, marketing strategies, growth forecasts, and investments in strategic account management. However, the true test of leadership lies in your ability to anticipate challenges—identifying early signals of dissatisfaction, diminished value, or service concerns before they escalate. Traditional metrics like Net Promoter Scores or basic satisfaction surveys often miss these nuances. Instead, proactive engagement and deep listening are your best tools for uncovering hidden risks and opportunities. By staying attuned to these signals, you position your organization to respond swiftly, adapt intelligently, and sustain competitive advantage.

For Sales Enablement

A clear understanding of what customers truly value is the linchpin of effective sales enablement. This knowledge helps you distinguish what's working from what isn't, and pinpoints where to focus your efforts—be it in support initiatives, talent acquisition, education, content development, or strategic priorities. Regularly reviewing customer feedback, segmented by account category, provides some of the richest insights available for refining your enablement programs. Treat this feedback as your compass: It will guide your team toward the most impactful actions and ensure that your resources are invested where they will drive the greatest results. Embrace this continuous learning process, and you'll empower your sales teams to exceed expectations and deliver exceptional value.

For Commercial Insight Strategists

Knowing what customer stakeholders think is valuable input for mapping out desired outcomes for individual opportunities and understanding how individuals influence decisions. Actively mining stakeholder health check responses and incorporating them into stakeholder profiles and buying center maps is highly beneficial. You will also take the lead in generating reports to show trends in how stakeholder perceptions—by individual, department, or level—change. Monitoring these changes is important for identifying both risks and opportunities. Consider health check monitoring a primary, valuable input for all aspects of managing account insights.

Ready to move beyond simply recognizing opportunities and risks? You've already laid the groundwork by mapping out your relationships and clarifying your direction. Now imagine what's possible when you harness the power of your network—not just the people you know, but the influence they carry and the connections they can open for you.

Are you curious about how to transform your existing relationships into a dynamic web of influence? Ready to discover how a single connection can spark a chain reaction of trust, credibility, and growth within your accounts? In the next chapter, we'll dive into the art and science of building your "spheres of influence," unlocking the strategies that top account sellers use to expand their reach and impact.

CHAPTER 5:

MAKING CONNECTIONS

've learned that nothing accelerates credibility with new executives like a meaningful referral or word-of-mouth endorsement from respected peers or industry experts. In today's information-saturated world, the relationships we cultivate and the influence those connections can exert are our greatest assets in growing enterprise accounts. Connections are not just useful; they are essential. Mastering the art of leveraging both direct relationships and the extended networks of those you know has become a core discipline, especially in our interconnected, social media–driven environment.

Your goal is to intentionally expand your network within and across accounts. Even if your current relationships are limited, they provide a foundation of familiarity and trust that you can build upon. Recognize that trust is the currency of new relationships, and every connection you make accelerates the process of earning it. Establishing trust quickly with new contacts is not just advantageous; it's essential for building the strong, resilient relationships that drive long-term success.

I've noticed that many account sellers overlook the power of relationship connections as the primary path to account expansion. Yet referrals remain the single most influential factor in opening doors and winning new advocates. That's why you need to take the time to map out your connections—from existing relationships to those you aspire to build. By understanding the nature of each connection, you can approach new stakeholders thoughtfully and leverage introductions most effectively.

Don't see this as cold outreach. Instead, focus on warm, intentional connection building that transforms your network into a "commercial community," with yourself at the center, creating value for everyone involved. Strive to see the bigger picture: becoming a connector and a trusted resource who brings people together. When you approach your relationships deliberately and strategically, the network effect multiplies your impact and accelerates the growth of your accounts. The power of a well-managed network cannot be underestimated—it's the engine behind exponential relationship growth.

UNDERSTANDING THE NETWORK EFFECT IN ENTERPRISE ACCOUNT MANAGEMENT

The **network effect** is a foundational business principle: As more individuals engage with a product or service, its value grows exponentially. In the context of enterprise account relationships, this means that the more stakeholders who interact with your organization, the greater the mutual value created. This dynamic is not just theoretical; it is a critical driver of account growth and expansion, closely tied to the relationship rhythms discussed previously.

Metcalfe's law further clarifies this principle, stating that the value of a network is proportional to the square of the number of connected users. In practical terms, even a modest increase in engaged stakeholders can dramatically amplify value. As accounts grow and more participants become involved, the network effect accelerates growth, strengthens retention, and raises barriers to entry for competitors. As an account leader, you are uniquely positioned and responsible for fostering this effect within and across your accounts. The opportunity is significant; let's examine how you can strategically leverage the network effect to your advantage.

MAPPING CONNECTIONS: THE FOUNDATION OF ACCOUNT EXPANSION

People are connected in myriad ways—through education, shared ideas, professional roles, backgrounds, skills, interests, employers, associations, events, geography, industry, family, and social circles. Forrester stats that:

> *Gen Z and millennial buyers now make up more than two-thirds of all B2B buyers and indicated that ten or more people from outside their organization were involved in their purchase decision.*[15]

The strength and nature of these connections vary, but your task is clear: Diligently explore, identify, and record the connections your key account stakeholders possess. Utilize

---

15 Barry Vasudevan, "Predictions 2025: Younger Business Buyers and GenAI Will Upend the Status Quo," *Forrester* (blog), October 24, 2024, https://www.forrester.com/blogs/predictions-2025-business-buyers/.

connection-mapping tools to systematically chart these relationships, ensuring each connection type is clearly labeled and
documented.

Why Is This Important?

Because the depth and breadth of these connections directly
influence your ability to expand and solidify account relationships.
The more connection points you uncover and track, the stronger
your foundation for growth.

Types of Connections: Building Blocks of Influence

To maximize account expansion, focus on the following core
connection types:

- **Role Connections:** Stakeholders in similar roles often
  share common objectives and experiences. By connecting these individuals across geographies, subsidiaries,
  or business units, you enable peer advocacy and amplify
  positive outcomes.
- **Shared Perspectives:** Executives frequently express their
  viewpoints in public forums. Identifying shared perspectives—on strategy, success criteria, or vision—creates a
  powerful basis for connection and collaboration. Don't
  overlook the influence of shared passions, whether professional or personal.
- **Association Connections:** Membership in professional
  or personal associations (industry groups, charities, clubs,
  educational or public service organizations) offers another
  avenue for relationship building. Shared memberships can
  open doors to new conversations.

- **Location Connections:** Shared geography, whether current or past, fosters affinity and common experiences. Leverage these bonds—formed through local schools, restaurants, and activities—to deepen relationships.
- **College or University Connections:** Alumni ties are often strong and enduring. Connections formed during these formative years, especially around school traditions or sports, can be especially influential.
- **Board Connections:** Board memberships offer unique opportunities for introductions and endorsements. A recommendation from a board member can be a decisive factor in gaining access.
- **Prior Employer Connections:** Shared work history, particularly during overlapping tenures, creates bonds rooted in common experience—positive or negative. These connections often translate into trust and mutual understanding.
- **Event Connections:** Recent or upcoming events provide timely opportunities to engage stakeholders. Use these occasions to establish or strengthen connections, leveraging shared experiences or interests.
- **Partner Connections:** Common relationships with suppliers, consultants, analysts, or other partners can be surprisingly fertile ground for new introductions and collaborations.

Becoming a Connection Hub: Your Role as a Connector

Position yourself as the connection hub—the individual who creates value by linking people with shared interests, experiences, and objectives. When you consistently bring stakeholders together, you become an indispensable resource, trusted for your ability to facilitate introductions, share insights, and foster meaningful relationships. You'll be surprised at how well this works.

The most successful account growth leaders excel at this. They cultivate networks that rely on them for access, information, and opportunity. This approach is not only effective, but it also makes your role more engaging and enjoyable. Remember, this is not about selling; it's about connecting people to value, ideas, and each other.

BUILDING A SOCIAL NETWORK MAP: YOUR STRATEGIC ADVANTAGE

Mapping the relationships within and around your enterprise accounts is not just a best practice; it's a critical strategy for unlocking hidden opportunities and accelerating growth. Traditional organizational charts and buyer networks provide only a partial view, often missing the complex web of connections that span geographies, subsidiaries, and business units. To truly understand and influence decision-making, you must look beyond formal structures and capture the informal, yet powerful, "tribal knowledge" that drives real business outcomes.

Finding Connections to Reach the Executive Buyer

5.1: Social Network Map

Think of yourself as a connector, someone who sees the bigger picture and understands how people are linked across the ecosystem. Just as investigators piece together relationships to solve cases, you can use social network mapping to identify key influencers, gatekeepers, and potential allies. This approach empowers you to do the following:

- **Uncover Hidden Pathways:** Reveal connections that are not visible on org charts but are crucial for account expansion.
- **Facilitate Warm Introductions:** Leverage existing relationships to gain access to new stakeholders, making your outreach more effective and credible.
- **Accelerate Trust Building:** Use shared connections to establish rapport quickly and lay the groundwork for lasting business relationships.
- **Outpace the Competition:** Move beyond cold prospecting by activating your network effect, giving you a significant head start over others.

Be proactive and intentional in mapping these relationships. Regularly update your diagrams as new connections emerge and business dynamics evolve. Remember, your ability to expand account relationships hinges on your willingness to see yourself as a connector and to harness the full potential of your network. By doing so, you position yourself and your organization for sustained success.

MOVING BEYOND RELATIONSHIPS: BUILDING SPHERES OF INFLUENCE

In today's dynamic business environment, your ability to influence is paramount. Influence is not merely about forming new relationships; it is about cultivating trust through relevance and sustained engagement. Trust, in turn, is the foundation upon which all meaningful business outcomes are built.

To maximize your impact, think strategically about your network. Don't limit yourself to individual connections. Focus on building broader spheres of influence. These spheres allow you to coordinate multiple influencing factors, increasing your ability to guide decision-makers toward mutually beneficial outcomes.

Sphere of Influence

5.2: Sphere of Influence

At the heart of your sphere of influence is the key stakeholder—the central figure, or "hero," as outlined in chapter 2. Surrounding this stakeholder is a network of buyers and influencers who collectively shape every decision. It is essential to map these relationships carefully, identifying not only direct contacts but also the secondary influencers who can advocate on your behalf.

Leverage your connections to move beyond simple introductions. Use your network to discover shared needs and to showcase how others have successfully addressed similar challenges with your company's solutions. When you bring trusted influencers into the conversation, you enhance your credibility and position yourself as a reliable partner.

However, proceed with caution: Influence must be earned, not assumed. Authenticity and relevance are critical. Stakeholders will quickly disengage if they sense manipulation or insincerity.

Remember, your goal is not just to win a deal, but to build a legacy of trust and value. By strategically expanding your sphere of influence and aligning your resources with stakeholders' needs, you empower both yourself and your clients to achieve lasting success. Take the initiative, remain vigilant, and always act with integrity—your influence will grow, and so will your impact.

BUILDING A COMMERCIAL COMMUNITY

Successful teams consistently demonstrate that account expansion is not merely a function of sales tactics but the result of strategic collaboration across divisions and the cultivation of deep, trusted relationships. In enterprise environments where organizations invest heavily in solving complex challenges, those who establish credibility and trust with multiple stakeholders can accelerate growth and unlock new opportunities across diverse buying centers. See chapter 11 for more details on how to set up a digital environment for a commercial community.

As a connector, your ability to develop internal networks, foster spheres of influence, and serve as a hub of trust is invaluable.

This foundation enables you to move beyond transactional relationships and build what we call a "commercial community." A commercial community transcends individual accounts, uniting professionals with shared interests, typically within a specific industry. Most industries are shaped by an ecosystem dominated by a handful of leading firms, where executives often spend their entire careers. By bringing together influential professionals from these top organizations and positioning yourself as the central facilitator, you create a powerful network that recognizes you as an essential part of the ecosystem.

Commercial Community

5.3: Commercial Community

However, this level of influence comes with responsibility. It is critical to maintain integrity and prioritize the collective interests of the community. Misusing your position or neglecting the needs of the group can quickly erode trust and diminish your standing.

Elevating yourself and your organization to this level allows you to shape industry conversations, influence agendas, and forge enduring relationships that outlast any single transaction. The most effective consultants excel in this arena by sharing insights, fostering collaboration, and driving innovation for the benefit of the entire community. They become recognized thought leaders and indispensable contributors to their networks.

In our final chapter, we will explore practical strategies for building and nurturing commercial communities in a digital landscape, one account at a time. Remember, the true strength of any community lies in the quality of its connections. By becoming the hub that unites and empowers others, you position yourself and your company for sustained influence and long-term success.

ROLE BY ROLE: STEPS TO SHAPE THE FUTURE
For Account Teams

Establishing a commercial community is a strategic move that streamlines interactions and deepens account relationships. By fostering an environment where shared objectives unite stakeholders, account teams gain a robust platform for advancing both vendor and customer goals. In today's fast-paced business landscape, these communities are essential for reducing friction, accelerating collaboration, and leveraging the positive experiences of existing stakeholders. Embrace this approach to unlock new opportunities and drive sustained relationship growth.

For Marketing

Understanding the intricate perspectives of enterprise accounts and their diverse stakeholders is invaluable. In a marketplace where agility and alignment with evolving customer needs are crucial for differentiation, commercial communities offer a decisive advantage. They break down organizational silos, encourage cross-functional collaboration, and provide continuous access to rich insights and research. Marketers can harness these communities to develop more relevant messaging and design impactful account expansion programs that directly support account teams.

For Management

Navigating the complex web of stakeholders within each account is a persistent challenge for management. With executives typically dedicating only a small fraction of their time to direct customer engagement, commercial communities offer a practical solution. These platforms facilitate seamless information exchange with key stakeholders, ensuring management has a comprehensive view of account dynamics. By fostering shared goals and unified objectives across vendor and customer teams, management can drive alignment and advance strategic agendas with greater efficiency.

For Sales Enablement

Commercial communities lay the groundwork for enhanced account team collaboration and enable efficient digital communication between vendors and customers. This approach reduces operational friction, boosts efficiency, and strengthens both relationship quality and customer retention. For sales enablement professionals, the focus should shift from deploying isolated tools to empowering account teams to map and cultivate these communities. Doing so fosters systematic growth and positions the organization for sustained success.

For Commercial Insight Strategists

Mapping connections between executive buyers within an account is highly valuable for account expansion. One of the first steps is to identify current stakeholders, explore potential new expansion areas, and examine the existing connections. This provides account executives with immediate outreach opportunities by leveraging their relationships within the account. Additionally, you can analyze the connections between account stakeholders across the industry. Create a commercial community map that outlines these connections across multiple accounts, with your company at the center. Use this map to introduce, align, and share information within the community to gain central leverage.

You've seen how building commercial communities can unlock new levels of collaboration and value—but what happens when you take that spirit of connection and aim it directly at the heart of your biggest opportunities? Let's turn the page and discover how you can move beyond the obvious, reframe the conversation, and start selling not just what's in front of you, but what's truly possible. Ready to rethink what your deals could become? The next chapter is where the real transformation begins.

CHAPTER 6:

SELLING WHAT'S POSSIBLE

Throughout my career, I've learned that the most successful account teams are those that don't sell just what's available; they sell what's possible. By intentionally aligning capabilities, resources, and value messaging to the unique needs of each stakeholder within a customer's buying network, account teams can unlock and accelerate larger, more impactful deals. The way we frame and approach a customer's problem directly influences the scope and value of the solutions we propose and, ultimately, the scale of the opportunities we capture. It's essential to invest the time to evaluate how well our solutions align with the customer's broader objectives, as this can significantly expand the potential of every engagement.

However, I've seen firsthand how easy it is to fall into the trap of offering quick, familiar solutions. The pressure to push specific products or services often leads us to recommend what's readily at hand rather than exploring the full spectrum of possibilities. This approach not only limits the value we deliver to our clients but also leaves substantial revenue untapped. Clients are looking for partners who can help them achieve their most ambitious outcomes, not just address immediate needs with point solutions.

Settling for a deal that's a fraction of its potential, many times smaller, often stems from a lack of credibility with senior stakeholders or from not framing the opportunity at the right level. Most executives expect us to identify the largest possible impact and work backward to engineer a solution that delivers it. Yet, without a deliberate process to align our solutions to the highest level of business impact, we risk missing out on transformative opportunities that are bigger in scope and more valuable to customers.

A powerful illustration of this challenge comes from the well-known "invisible gorilla experiment" by Christopher Chabris and Daniel Simons. Participants, focused on counting basketball passes between a group of people in a circle, often failed to notice a person in a gorilla suit walking right through the middle of the circle! This phenomenon, known as inattentional blindness, highlights how easily we can overlook critical opportunities when our attention is narrowly focused. In sales, we're often so absorbed in managing the tactical exchanges between stakeholders and internal teams that we miss the bigger, game-changing possibilities right in front of us.

I urge you to step back and deliberately assess not just how many opportunities exist, but how big each opportunity can truly be. By broadening our perspective and adopting a structured, analytical approach to solution design, we can consistently deliver greater economic value to our clients and drive larger, more strategic deals for our organization. Let's challenge ourselves to see beyond the obvious, avoid the pitfalls of inattentional blindness, and become the trusted advisers our clients expect, capable of envisioning and delivering what's truly possible.

LAYING THE FOUNDATION FOR BIGGER DEALS: THE POWER OF COMMERCIAL RELATIONSHIPS

As discussed in chapter 4, understanding the rhythm of your client relationships is fundamental to securing larger, more strategic deals. We explored four key relationship types: responsive, accommodative, prescriptive, and performance. Each represents a distinct level of trust and influence, and your ability to accurately assess and intentionally evolve these relationships is critical.

It is important to recognize that without a solid relationship foundation, your recommendations, no matter how valuable, may lack credibility and fail to gain traction. Only by building trust and demonstrating reliability can you position yourself to propose solutions that are broader in scope, higher in value, and longer in duration.

Approach each relationship with purpose. Take deliberate steps to advance it, aligning your actions with the client's needs and expectations. This strategic progression not only opens the door to higher-margin opportunities but also cements your role as a trusted adviser.

Remember, Selling What's Possible™ begins with the relationships you cultivate today. Invest in them thoughtfully, and you will unlock the potential for transformative, long-term success.

COMMON CAUSES OF DEAL SCOPE AND SIZE MISALIGNMENT

Successfully driving growth within key accounts hinges on more than a compelling value proposition. Lasting account expansion is the result of a nuanced approach that identifies and engages the right stakeholders, tailors messaging to their unique priorities, and steadily earns trust throughout the organization. However, even experienced professionals can fall into traps that hinder progress and credibility. Avoiding critical missteps at each stage of the account management process is essential to elevating your influence and achieving measurable results.

1. Engaging at the Wrong Stakeholder Level

A frequent pitfall in account selling is misaligning your pitch with the stakeholder's level of authority. Senior executives are focused on strategic, high-impact initiatives, not transactional, low-value purchases. Presenting such offers to executive leaders not only wastes their time but can also result in being permanently excluded from future opportunities at that level. Remember: Elevating your pitch is only effective when the solution matches the executive's mandate and priorities. Misalignment is all too common, as evidenced by the flood of irrelevant outreach executives receive daily.

2. Failing to Understand Stakeholder Roles and Responsibilities

Another common oversight is neglecting to research and understand the specific needs of each stakeholder involved in the decision-making process. Stakeholder priorities vary by role and organizational level. To achieve true alignment, it is essential to identify the "job to be done" for each role and clearly demonstrate how your solution advances their objectives. This targeted approach is critical for building relevance and credibility.

3. Lacking Credibility with Key Stakeholders

Credibility is not given; it is earned. It can be built through peer recommendations, successful case studies, demonstrated business acumen, and a strong company reputation. Additionally, internal references from within the buyer's organization can be particularly influential. Purposeful engagement with lower-tier stakeholders, while keeping a long-term vision for larger deals, can also lay the groundwork for future success. Consider how you can strategically leverage these assets to strengthen your position.

Strategic Alignment: The Foundation for Account Growth

Misalignments frequently arise—even early on—within established account relationships, making it essential to thoroughly understand the customer's growth strategy and relevant timelines before attempting to expand your influence. Corporate visions research shows that:

> *"Only 45 percent of sellers and buyers align on the core problem after discovery conversations, and in lost deals, that number plummets to 23 percent."*[16]

Advancing without this clarity leads to undefined objectives and hinders the execution of a focused, scalable growth plan. To

---

16 Tim Riesterer, "Unlearning Discovery: New Research Shows Why B2B Sales Must Evolve," *Corporate Visions*, accessed September 21, 2025, https://corporatevisions.com/blog/b2b-sales-discovery-research/.

achieve lasting success in account selling, consistently align your approach with the right stakeholders, appreciate their distinct roles and priorities, and build credibility at every organizational level. By anchoring your efforts in a well-defined account growth strategy, you not only avoid common pitfalls but also drive meaningful, sustainable results. Stay strategic, remain adaptable, and always focus on delivering value where it counts most. This is the path to enduring client partnerships and business growth.

THE STRATEGIC VALUE OF BIGGER DEAL SIZES

In the rapidly evolving, AI-driven business landscape, sales organizations face unprecedented pressure to increase efficiency and reduce costs. Automation streamlines routine sales transactions, often at the expense of account relationships. While this shift can lower operational expenses, it also compresses margins and threatens the personalized service that distinguishes high-performing sales teams.

Why Bigger Deals Matter More Than Ever

Focusing solely on increasing the number of smaller, automated transactions exposes organizations to a race to the bottom on price and profitability. As AI commoditizes the sales process, sustainable growth depends on a different approach: prioritizing larger, higher-value deals that deliver superior margins.

- **Strategic Engagement:** Larger deals typically arise from deeper, consultative relationships where sales professionals act as trusted advisers. These engagements enable sellers to collaborate with clients on solving complex problems or cocreating innovative solutions.

- **Margin Protection:** High-value contracts are less susceptible to margin erosion, as they are built on unique value propositions rather than price competition alone.
- **Enterprise Focus:** Working with enterprise accounts—those with the greatest spending power—positions organizations to drive high value outcomes for both parties, fostering long-term loyalty and mutual growth.

Navigating the Shift

To thrive in this environment, account teams must elevate their role within client organizations. This means moving beyond transactional selling and aligning with stakeholders who influence strategic decisions.

- **Develop Domain Expertise:** Invest in building knowledge and skills that enable your team to address complex client challenges and identify new growth opportunities.
- **Cultivate Strategic Relationships:** Proactively engage with decision-makers and influencers at higher organizational levels to uncover and shape enterprise-wide initiatives.
- **Align Growth Strategies:** Focus on creating value that extends beyond the immediate sale, positioning your organization as an indispensable partner in the client's long-term success.

Now is the time to embrace a growth strategy centered on winning bigger, higher-margin deals. By focusing on strategic partnerships and delivering exceptional value, your team can secure profitable growth and a sustainable competitive advantage. Elevate your approach, deepen your client relationships, and lead your organization into a future defined by meaningful, high-impact sales success.

ACCOUNT GROWTH STRATEGY: A GUIDE FOR ENTERPRISE SUCCESS

Defining an Account Growth Strategy

In today's rapidly evolving business environment, a robust account growth strategy is essential, especially for enterprise clients whose investments drive significant value. The focus must shift from chasing high-volume, low-margin transactions (increasingly automated by AI) to cultivating larger, higher-value, and higher-margin deals. This approach not only offsets the commoditization of transactional sales but also positions your team to deliver lasting impact.

Key Principles for Effective Growth

- **Stakeholder-Centric Approach:** Successful account teams start by deeply understanding the unique needs and challenges of each stakeholder. By listening first and aligning solutions to these specific requirements, teams can deliver tailored value that resonates.
- **Solution Alignment:** Work backward from the client's objectives, ensuring that every proposed capability, resource, and message is outcome-driven and relevant.
- **Value-Based Relationships:** Prioritize building relationships that go beyond transactional exchanges, focusing on delivering measurable business outcomes and establishing trust.

The Rewards of Strategic Execution

Proactively implementing a growth strategy centered on value unlocks a cycle of success: as you deliver bigger, higher-margin deals, you naturally uncover additional opportunities within the client's organization. Enterprise accounts are complex, with multifaceted needs—addressing these holistically creates a foundation for enduring partnerships.

Sustaining Competitive Advantage

Achieving high-value relationships is only the beginning. The real advantage comes from maintaining and strengthening these connections over time, which creates a sustainable competitive edge:

- **Visualize Your Strategy:** Clearly map out your growth objectives and the path to achieve them.
- **Prioritize Opportunities:** Focus efforts where they will have the greatest impact, aligning resources to the most promising areas.
- **Lead with Purpose:** Guide your account teams with clarity and conviction, ensuring disciplined execution at every stage.

Lead with Vision, Win with Value

Adopting this strategic mindset requires discipline, empathy, and a commitment to delivering genuine value. The journey may be complex, but the rewards—stronger relationships, larger deals, and a defensible market position—are well worth the effort. Lead with vision, act with intent, and inspire your teams to pursue growth that benefits both your clients and your organization.

VISUALIZING YOUR GROWTH STRATEGY: A DASHBOARD APPROACH

A well-structured dashboard is an essential tool for steering your account growth strategy toward higher-value relationships. By leveraging a straightforward 2x2 framework, you can clearly map where your current contracts sit and identify opportunities for advancement. This approach not only clarifies your present position but also illuminates the path to greater revenue and margin.

Understanding Contract Growth Dimensions

Effective account growth is driven by evolving contracts, making them larger, more valuable, and more strategic. As an account seller, your mandate is to proactively expand the scope and impact of every contract. There are five key contract growth strategies to consider:

- **Enrich:** Add value to existing contracts by introducing additional options or features.
- **Enlarge:** Broaden contracts to encompass more solutions, increasing their overall scope.
- **Extend:** Secure new funding by extending contracts into other budgets managed by different buying centers.
- **Expand:** Introduce entirely new solutions to new buying centers, opening fresh avenues for growth.
- **Elevate:** Transform contracts into comprehensive, enterprise-level solutions that deliver greater strategic value.

Using the Growth Strategy Dashboard

Before diving into specific examples of each growth strategy, take a moment to familiarize yourself with the dashboard view.

Visualize each contract component as a "pin" on this dashboard, representing its current contribution to revenue. By mapping your contracts, you gain immediate insight into your business's present state.

For many organizations, most revenue is concentrated in the lower left quadrant—transactional contracts focused on repeat product sales to new customers. This pattern reflects a responsive, product-centric rhythm that, while effective, may limit long-term account value.

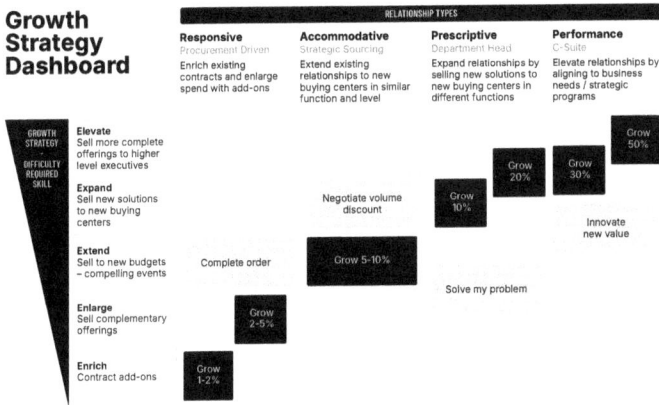

Growth Strategy Dashboard

	RELATIONSHIP TYPES			
	Responsive Procurement Driven	**Accommodative** Strategic Sourcing	**Prescriptive** Department Head	**Performance** C-Suite
	Enrich existing contracts and enlarge spend with add-ons	Extend existing relationships to new buying centers in similar function and level	Expand relationships by selling new solutions to new buying centers in different functions	Elevate relationships by aligning to business needs / strategic programs

GROWTH STRATEGY · DIFFICULTY REQUIRED SKILL

Elevate Sell more complete offerings to higher level executives

Expand Sell new solutions to new buying centers

Extend Sell to new budgets – compelling events

Enlarge Sell complementary offerings

Enrich Contract add-ons

Complete order — Negotiate volume discount — Grow 5-10% — Grow 10% — Grow 20% — Grow 30% — Grow 50%

Solve my problem — Innovate new value

Grow 2-5% — Grow 1-2%

6.1: Growth Strategy Dashboard

It is crucial to recognize that remaining in transactional territory can restrict growth potential. Use your dashboard not just as a record, but as a springboard for strategic action. Challenge yourself and your team to move contracts up and to the right toward more expansive, integrated, and value-driven relationships.

Begin by mapping your pipeline opportunities onto the dashboard. Assess the current position of your pipeline: Are your opportunities concentrated at higher relationship levels, such as department heads or the C-suite? This analysis is the cornerstone of your growth strategy, targeting more opportunities where relationships are strongest and most strategic.

To sharpen your approach, consider the following critical questions:

- Do you fully understand the needs and opportunities at the department head or C-suite level?
- What is the nature and pattern of these needs?
- Can your organization deliver solutions that genuinely address these requirements?
- How do your offerings compare to competitive alternatives?
- What is the volume and variety of these needs?
- Where do your strengths align with the needs of these stakeholders?
- Do you have established connections with department heads, C-suite leaders, and their peers?
- What measurable business impact can you deliver, and will those outcomes meet or exceed expectations?
- What is the budget allocation within these key buying centers?
- Who are the decision-makers and influencers at these levels?

While there are numerous opportunities to pursue, it is essential to prioritize those with the greatest potential for impact and margin. Before setting priorities, examine what high-value, prescriptive, and performance-based deals look like in your context.

Prescriptive Rhythm deals focus on solving specific problems with tailored capabilities. This is the domain of department leaders and their teams. Your organization likely has successful examples of these engagements. However, timing is increasingly critical: Being first to identify and address emerging needs, in partnership with your customer, is a decisive advantage. Pre-intent signals, as discussed in chapter 1, offer early indicators of these opportunities. These customer-triggered growth events are frequent and create a consistent pipeline for proactive engagement.

Approach this process with discipline and curiosity. By systematically analyzing your pipeline and aligning your solutions to the highest-value relationships, you position your team for sustainable, high-impact growth. Seize the initiative. Prioritize strategically, act decisively, and lead the conversation at the top levels of your customers' organizations.

Growth Strategy Dashboard

RELATIONSHIP TYPES

	Responsive Procurement Driven	Accommodative Strategic Sourcing	Prescriptive Department Head	Performance C-Suite
	Enrich existing contracts and enlarge spend with add-ons	Extend existing relationships to new buying centers in similar function and level	Expand relationships by selling new solutions to new buying centers in different functions	Elevate relationships by aligning to business needs / strategic programs

GROWTH STRATEGY — DIFFICULTY REQUIRED SKILL

Elevate Sell more complete offerings to higher level executives — OPPORTUNITY

Expand Sell new solutions to new buying centers — OPPORTUNITY OPPORTUNITY

Extend Sell to new budgets – compelling events — OPPORTUNITY OPPORTUNITY OPPORTUNITY OPPORTUNITY

Enlarge Sell complementary offerings — OPPORTUNITY

Enrich Contract add-ons

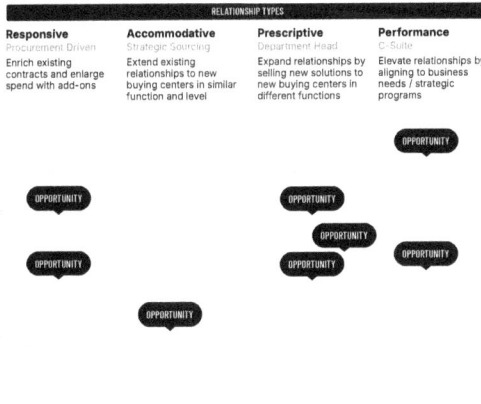

6.2: Plotting Customer-Triggered Growth Opportunities

Consistently demonstrating your expertise at critical moments by offering targeted, actionable insights significantly increases your likelihood of success. By proactively addressing clients' specific challenges or objectives with prescriptive guidance, you position yourself as a trusted adviser and dramatically improve your win rates.

CASE IN POINT—HOW THE SAME NEED CREATED TWO DIFFERENTLY SIZED OPPORTUNITIES

This client was navigating the complexities of enabling its advertising customers to select addressable audiences for executing ad campaigns targeting specific households versus media programs across varied markets. This was a new offering they wanted to enable.

Prescriptive Rhythm Opportunity: Strategic Innovation

At first we partnered with the ad sales division's leadership to evaluate advanced indexing solutions. By implementing a new cross-referenced audience index, we empowered their customers to identify, segment, and activate audiences more efficiently, thereby streamlining campaign planning, targeting, and pricing across regions.

This initiative led to a contract expansion with a new division, underpinned by a unique solution that had no direct market equivalent. Pricing was value based, reflecting the transformative impact delivered to the client. The absence of comparable alternatives allowed us to justify our approach and secure buy-in at a premium, demonstrating the strategic advantage of tailored, insight-driven solutions.

Performance Rhythm Opportunity: Unlocking Transformational Growth

The team involved took this opportunity to the next level. Building on this momentum, we introduced the same client to a groundbreaking approach in advanced advertising, moving from geographically indexed targeting based on "zones" to individualized household engagement. Leveraging set-top box data, we enabled precise, device-level ad delivery tailored to granular household characteristics.

Our team's research and vision for the future of advertising were compelling enough to engage the company's president directly, with executive coaching focused solely on business impact. The mandate was clear: Deliver over $1 billion in value within three years, or the initiative would not proceed. Our credibility and clarity of purpose secured immediate executive approval and a dedicated project leader.

Within a single meeting, we launched an innovation now shaping the future of TV audience buying and selling—driving industry-wide change and measurable value creation.

Win Big by Solving What Others Miss

These examples demonstrate a critical lesson: Transformative, high-value deals are often easier to win when your solution is unique, addresses a pressing need, and faces little competition. Value-based pricing becomes not only possible but a clear alternative acceptable to the client.

However, success demands a strategic growth mindset and disciplined execution. Identifying unmet needs and crafting bespoke solutions is the cornerstone of insight-driven selling, the future of our industry.

PRIORITIZING OPPORTUNITIES TO PURSUE

Effective opportunity prioritization begins with a rigorous Bluespace analysis, as outlined in chapter 1. This analysis is not just a theoretical exercise. It is the foundation for identifying and evaluating the most promising opportunities within your account portfolio. Once you have thoroughly assessed the signals indicating your account's needs, it is essential to analyze the "altitude" of the stakeholders involved in the buying process.

Ask yourself the following:

- Where are these needs originating within the organization?
- Are department heads or C-level executives engaged or likely to be involved?
- Who are these key stakeholders, and do we have established relationships with them?
- Have we successfully addressed similar challenges in the past?
- How urgent is the need? Does it present a significant threat that demands immediate attention?
- Are we already solving related issues within this account?

As you narrow your focus to the top ten opportunities, prioritize those that reach the department head or C-suite level. These are typically the most strategic and impactful, warranting careful evaluation and pursuit.

However, proceed with caution: Engaging executive buyers demands preparation, insight, and a tailored approach. Success is not guaranteed simply by targeting high-level stakeholders. To maximize your chances, you must secure initial meetings, build a shared vision of success, and systematically advance the opportunity toward a closed deal.

Remember, prioritizing opportunities is both an art and a science. By applying disciplined analysis and strategic judgment, you position yourself and your organization to win the deals that matter most. Stay focused, stay curious, and always aim higher. The opportunities you choose to pursue today will define your success tomorrow.

ENGINEERING WHERE TO LAND A DEAL

To strategically position your deal for success, begin by clearly identifying the desired deal type, aligning it with your contract objectives and the cadence of your relationship with the client. Next, apply a structured checklist to systematically evaluate any gaps between your current standing and the requirements for success at this level. If you determine that these gaps can be closed, the opportunity is viable and should be pursued at this organizational level. If not, consider recalibrating your approach to target a more transactional deal at a lower level, while maintaining a long-term goal of elevating future engagements.

Always aim to close capability gaps and advance toward higher-value deals. This approach not only maximizes immediate opportunities but also builds organizational credibility and influence over time.

Growth Strategy Dashboard

	Responsive	Accommodative	Prescriptive	Performance
Elevate Sell more complete offerings to higher level executives				●
Expand Sell new solutions to new buying centers			⊘	
Extend Sell to new budgets –compelling events		● ●		
Enlarge Sell complementary offerings	●			
Enrich Contract add-ons				

Gap Checklist

☐

☑

☑

☑

☑

☑

6.3: Deal Gap Checklist

CHECKLIST CRITERIA: A STRATEGIC GUIDE TO OPPORTUNITY QUALIFICATION

Qualifying business opportunities requires more than intuition or a surface-level review. Success stems from a disciplined, criteria-driven approach that uncovers both obvious and hidden factors essential for winning and delivering value. Rigorous qualification helps prioritize where efforts and resources yield the greatest impact, while protecting your organization from costly missteps. To maximize win rates and resource effectiveness, examine each potential pursuit through the lens of stakeholder engagement, organizational capabilities, resourcing, outcome alignment, and client urgency. This structured method empowers you to focus on opportunities where your team can build trust, meet needs with confidence, and achieve results that matter.

1. Stakeholder Engagement

Assess your understanding of all key stakeholders, especially those who hold decision-making power. Leverage your opportunity report's buying center analysis to determine your ability to access and influence these VIPs. Gaining true access is not just about knowing names. It's about building trust and credibility with those who matter most.

2. Required Capabilities

Evaluate whether your organization possesses both the technical capabilities and the expertise needed to deliver a compelling solution. Align your strengths with the client's expectations and requirements. Ask yourself: Does your solution address their needs, or are you asking them to take a leap of faith? The risk must be minimized, and the fit must be unmistakable.

3. Necessary Resources

Confirm that you have the right team in place, including subject matter experts, from initial engagement through execution. Resource gaps can erode credibility and jeopardize delivery. Ensure your team's availability and commitment before proceeding.

4. Ability to Achieve Desired Outcomes

Demonstrate, with evidence, your capacity to deliver the business outcomes each stakeholder expects. Outcomes must be both significant and achievable within a reasonable time frame. If you can't prove impact, your proposal will not stand out.

5. Priority of the Need

Determine the urgency of the client's need. Is this initiative a top priority, or will it be lost among competing projects? Drawing from Paul Davison, Bryan Gray, Mike Rendel, and Jesse Laffen's insights in *The Priority Sale*, recognize that clients act swiftly when faced with a genuine threat, not just an aspiration. Your solution must address a critical challenge that demands immediate attention.

Qualify Ruthlessly, Focus Where You Can Win, and Make Every Opportunity Count

Approach each opportunity with discipline and strategic intent by leveraging this checklist as your critical evaluation tool. Begin by thoroughly assessing every opportunity against the established criteria, carefully identifying and documenting any gaps that may exist. Once gaps are recognized, proactively develop targeted strategies to address them before committing additional time or resources.

Be mindful: Advancing opportunities that do not meet these

standards can lead to wasted resources and erode your organization's credibility. Each pursuit demands significant investment, often translating to hundreds of thousands of dollars and extensive effort over many months. Therefore, it is essential to focus only on those opportunities where you are best positioned to win and deliver measurable value.

Remember, the most successful organizations set themselves apart by rigorously qualifying and prioritizing their pursuits. As highlighted by David Brock:

> *"Teams that consistently apply these best practices have achieved win rates approaching 95 percent."*[7]

Let this serve as your inspiration and benchmark. By embracing disciplined qualifications, you not only sharpen your competitive edge but also ensure your efforts drive meaningful results. Make every pursuit count and focus where you can truly make an impact.

EXECUTING YOUR ACCOUNT STRATEGY

With a robust process for evaluating, prioritizing, and activating high-value opportunities now established, it is essential to align your account plan with your overarching growth strategy. Step back and assess the full landscape of Bluespace opportunities at your disposal. Identify and prioritize your top ten Prescriptive and Performance opportunities. For each, develop a targeted action plan that ensures effective stakeholder engagement, fosters a shared vision, and strategically aligns your team's

---

17 "Great Problem Solvers Win by Shaping Decisions," *Selling What's Possible*, YouTube video, accessed September 21, 2025, https://www.youtube.com/watch?v=Eko_Tc7_XTQ.

capabilities, resources, and value propositions with the client's desired outcomes.

As you progress, remember that your role is not only to lead but also to serve as a trusted adviser. Maintain a proactive leadership stance, guiding clients toward solutions that are credible and tailored to their evolving needs. This approach demands diligence. Customers expect insightful, reliable guidance that is demonstrably aligned with their business objectives. Only when this alignment is achieved will you earn the trust required to repeatedly engage with key buying centers and unlock sustainable, exponential growth.

ROLE BY ROLE: STEPS TO SHAPE THE FUTURE
For Account Teams

Achieving ambitious growth targets demands a disciplined and strategic approach. Begin by mapping all potential opportunities on a growth strategy dashboard to visualize the landscape. Use a structured checklist to evaluate and prioritize deals, considering both the strength of client relationships and the specific objectives of each contract. This method not only clarifies where your highest-value opportunities lie but also ensures your efforts are focused on deals that maximize revenue and margin. Remember, success is built on deliberate planning and consistent execution. Don't leave growth to chance.

For Marketing

To drive impactful campaigns, it is essential to understand the relationship dynamics within your enterprise accounts. Analyze which types of connections are most prevalent and which solutions consistently resonate and deliver value. These insights should directly inform your marketing content, case study development, and targeted expansion initiatives. By aligning your messaging with proven success factors, you position your team to engage more effectively and accelerate growth within key accounts. Stay curious and data-driven. Your best campaigns are rooted in real client success.

For Management

Effective account growth starts with a structured, customer-centric strategy. Rigorously evaluate client signals and deal types pursued by your teams to gain a holistic view of where growth is most achievable. This approach empowers you to make informed investment decisions—whether in talent, industry sectors, or specific accounts—while maintaining clear metrics to track progress and effectiveness. Exercise both caution and ambition: Strategic focus and ongoing measurement are your best safeguards against wasted resources and missed opportunities.

For Sales Enablement

Enabling account teams with the right tools is critical for advancing complex, high-value deals. Go beyond basic prospecting materials—develop assets that address executive buyers' priorities, such as solution road maps, impact analyses, phased implementation plans, and resource requirements. Support account teams as they craft and refine their growth strategies, evaluate opportunities, and integrate these insights into actionable account plans. Investing in these enablement efforts not only accelerates deal progression but also builds lasting client trust and loyalty. Your guidance can be the catalyst that turns potential into performance.

For Commercial Insight Strategists

You will compare each opportunity to the targeted strike zone and the associated checklist for successful pursuit with your account executive and account team. This process helps determine the solution you will recommend. It is often a good idea to aim higher initially and then adjust downward to the next level; however, you need to be careful not to be so misaligned with what you can actually deliver and the customer's perception of you that you lose credibility. Point out instances where your firm has successfully achieved this before, and see if you can elevate your qualifications to align with the recommended solution scope.

Before we dive in, pause for a moment and ask yourself: Are your customer relationships truly fueling the growth you envision, or are you relying on a handful of heroic efforts to get across the finish line?

You've seen how trust and credibility set the stage for lasting partnerships—but what happens when the pressure is on, and your team must deliver results, quarter after quarter? The answer lies not in luck or last-minute heroics, but in a disciplined, collaborative approach to opportunity prioritization.

Ready to move beyond the status quo and transform the way your team drives account growth? Let's pull back the curtain on the real-world challenges—and surprising opportunities—that await in the next chapter.

CHAPTER 7:

PLANNING FOR BROAD AND DEEP RELATIONSHIPS

frequently find myself reminding CROs that top accounts are often the backbone of quarterly revenue. The response is almost always instant agreement. This reliance is not unique; it's the prevailing reality for most organizations. Throughout my tenure as a CRO, I have depended on highly engaged teams dedicated to our largest accounts. Their ability to secure crucial growth deals has consistently helped us achieve our targets. Yet I've noticed these "acts of heroism" tend to come from the same individuals time and again.

While every business leans on its largest accounts to deliver results, today's expectations for growth, especially within existing enterprise relationships, are higher than ever. However, relying on extraordinary efforts from a few is not a sustainable growth strategy. When you consider that account teams are composed largely of delivery resources, service professionals, and subject matter experts, it becomes clear that orchestrating daily account activity across the entire team is vital. I'm continually surprised by how account planning often centers on just the sellers, rather than leveraging the full expertise of the team when it comes to growth and retention strategies.

Managing the ongoing relationship with every customer stakeholder is a complex task. Teams must navigate a multitude of factors influencing the customer's business, alongside many unpredictable variables. Executing aggressive growth strategies under these conditions is daunting. The pace of change in our world means that account teams can quickly fall behind if they rely solely on manual processes. Staying relevant and proactively solving customer challenges at scale is nearly impossible without a more dynamic approach.

After more than three decades observing businesses across industries, I see the same pattern: Account planning remains a manual, sales-driven exercise, updated infrequently and focused narrowly on short-term product opportunities. These plans are often static documents, relegated to a shelf until the next review cycle. It's no wonder account leaders express frustration, questioning the value and impact of the process.

If we want to achieve sustainable growth, we must move beyond episodic heroics and embrace a more integrated, team-driven, and dynamic approach to account planning. By doing so, we empower our teams, deepen our client relationships, and position ourselves to meet and exceed the ever-increasing demands of today's marketplace.

RETHINKING ACCOUNT GROWTH TARGETS: A CALL FOR STRATEGIC PRECISION

In many organizations, annual revenue goals are met by distributing growth targets evenly across all existing accounts. While this method offers simplicity, it often masks a critical reality: Most companies lack the nuanced insights required to tailor growth objectives to each account's unique potential. Without this level of detail, setting effective, individualized targets becomes nearly impossible.

Gartner's recent findings underscore the risks of this traditional approach:

> *Sales organizations that cling to legacy pipeline generation and retention tactics consistently fall short of evolving buyer expectations and, as a result, miss their revenue goals.*[18]

The path forward demands a fundamental shift: aligning sales execution with deep buyer insights, robust account pipeline development, and a relentless focus on both retention and growth.

The Pitfalls of Outdated Account Planning

Today's account planning processes are often relics of the past—manual, complicated, and inflexible. These plans tend to be the following:

- **Overly Complex and Time-Consuming:** The process is labor-intensive, diverting valuable resources from growth-driving activities.

---

18 Gartner, "Winning Modern B2B Buyers," *Gartner*, April 2024, https://www.gartner.com/en/sales/trends/winning-modern-b2b-buyers.

- **Lacking Actionable Focus:** Plans are frequently static documents, not dynamic road maps for execution.
- **Siloed and Disconnected:** Operational barriers prevent seamless collaboration across the account team, stifling innovation and responsiveness.
- **Narrow in Perspective:** Without continuous input and adaptation, account plans offer only a limited, outdated view of opportunities and risks.

The result? Teams invest significant time in plans that rarely deliver meaningful value or drive account growth. The cost of the current state is 80 percent internal time versus 20 percent external time. Add up all account seller compensation and take 80 percent of that, and that is the idle time cost you currently have. Flipping this to 80 percent external time and 20 percent internal would result in an efficiency gain in terms of actual cost of 60 percent of the total account seller salaries and benefits. It's a big number.

Today's Account Planning Process

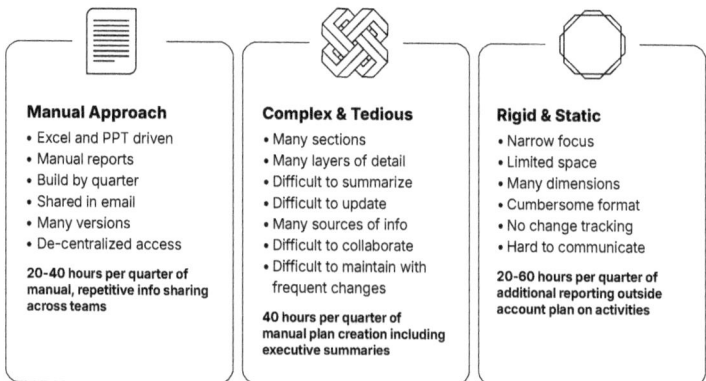

Manual Approach
- Excel and PPT driven
- Manual reports
- Build by quarter
- Shared in email
- Many versions
- De-centralized access

20-40 hours per quarter of manual, repetitive info sharing across teams

Complex & Tedious
- Many sections
- Many layers of detail
- Difficult to summarize
- Difficult to update
- Many sources of info
- Difficult to collaborate
- Difficult to maintain with frequent changes

40 hours per quarter of manual plan creation including executive summaries

Rigid & Static
- Narrow focus
- Limited space
- Many dimensions
- Cumbersome format
- No change tracking
- Hard to communicate

20-60 hours per quarter of additional reporting outside account plan on activities

7.1: Today's Account Planning Process is Manual and Time-Consuming

Redefining Account Plans

This reality prompts a crucial question: What is the real purpose of an account plan? It's time to move beyond routine, box-ticking exercises. Instead, let's embrace account planning as a strategic, collaborative, and actionable process—one that empowers teams to unlock growth, anticipate risks, and deliver exceptional value to clients.

By challenging the status quo and reimagining our approach, we can transform account planning from a burdensome task into a powerful catalyst for sustainable revenue growth and long-term customer success.

THE PURPOSE OF AN ACCOUNT PLAN

An account plan is a strategic tool designed to drive sustainable growth, strengthen client retention, and unlock new expansion opportunities. Its true value lies in providing a clear road map that aligns your team's efforts with the client's evolving needs and your organization's business objectives.

Today's Manual Account Plans vs Automated Account Plans

MANUAL ACCOUNT PLAN
Today's account plans are static and primarily look at a historical view of the account.

AUTOMATED ACCOUNT PLAN
Account plans should be forward-looking at the broad range growth opportunities and threats on the horizon through an interactive dashboard.

7.2: Today's Manual Account Plans vs. Automated Account Plans

The Current Reality and Its Limitations

Despite their intended purpose, many account plans today fall short of their strategic potential. Often, they exist as static PowerPoint decks capturing only a snapshot in time. Typically, these plans function as status reports for leadership, emphasizing revenue figures rather than a holistic view of the account.

This approach has several drawbacks:

- **Reactive vs. Proactive:** Plans are often retrospective, focusing on what has already occurred rather than anticipating future challenges and opportunities.
- **Inward-Facing:** They serve internal reporting needs more than client-centric strategy development.
- **Limited Scope:** Without a dynamic, comprehensive perspective, teams miss the broader context necessary for long-term success.

Ditch the Static—Make Account Planning Dynamic, Strategic, and Unstoppable

To truly unlock the power of account planning, organizations must move beyond static documents. Effective account plans should be living strategies—dynamic, forward looking, and deeply integrated with both client insights and market intelligence. By doing so, teams can drive sustained growth, deepen client relationships, and confidently navigate the complexities of today's business landscape.

Embrace account planning as a strategic discipline, not a routine task. This shift will empower your team to transform insights into action and deliver exceptional value to your clients.

DEFINING AN EFFECTIVE ACCOUNT PLAN

A robust account plan must go beyond simply mirroring CRM data or replicating traditional, static reports. While many online account planning tools integrate with CRM systems and link to email and calendars, they often fall short by requiring manual data entry and by presenting a fragmented, transactional view. This approach frequently overlooks the broader customer context, failing to capture the full picture across global regions, subsidiaries, departments, and ongoing initiatives.

The true purpose of account planning is to drive growth, retention, and broaden and deeper customer relationships. Yet too many plans remain backward-looking, focusing on past performance instead of anticipating future needs and opportunities. This limited perspective restricts the ability to identify actionable insights and hinders proactive engagement.

Modern organizations must embrace a new standard for account planning—one that leverages real-time data and the power of AI within a digital, interactive dashboard. By doing so, teams can do the following:

- Maintain a current, comprehensive view of each account as situations evolve.
- Foster collaboration across departments and regions.
- Enable timely, coordinated execution on opportunities and risks.

A dynamic, continuously updated account plan dashboard is essential for thriving in today's fast-paced business environment. It empowers teams to stay relevant, adapt quickly, and manage risk with greater precision.

Key Elements of a Modern Account Plan Dashboard

To maximize value, a dynamic account plan should include the following:

- **Real-Time Insights:** Integrate live data streams to reflect the latest developments, initiatives, and customer needs.
- **Holistic Customer Perspective:** Organize information across all business units, geographies, and projects to ensure nothing is overlooked.
- **Actionable Intelligence:** Highlight opportunities for growth, retention, and relationship expansion, not just historical outcomes.
- **Customer-Driven Alignment:** Systematically connect your organization's capabilities and resources to the customer's evolving priorities.
- **Continuous Collaboration:** Facilitate information sharing and coordinated action across the entire account team.

By adopting a modern, customer-centric approach, organizations can transform account planning from a static reporting exercise into a powerful engine for growth and partnership. The future belongs to those who act decisively, adapt rapidly, and align their strengths with the needs of their customers.

Dynamic Interactive Account Dashboards

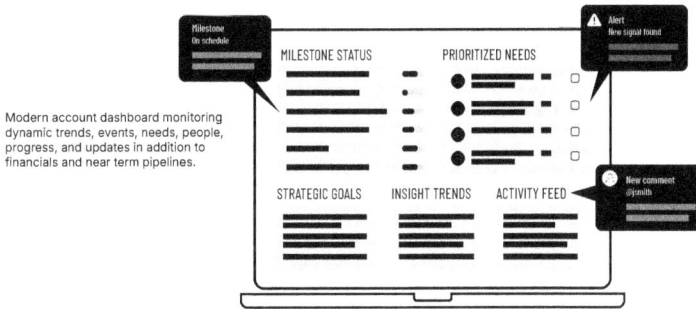

7.3: Dynamic Interactive Account Dashboards

UNLOCK GROWTH: EIGHT ESSENTIAL PILLARS TO MODERNIZE YOUR ACCOUNT PLANS AND BOOST BUSINESS IMPACT

Building an effective account plan demands more than sporadic insights and one-off customer check-ins. It requires a well-or-chestrated, customer-driven approach as its foundation. A modernized account plan thrives on systematic intelligence, daily prioritization, and unified collaboration, ensuring that your team remains agile and proactive in addressing evolving client needs. By methodically aligning actionable intelligence with prioritized growth opportunities, value alignment, and progress tracking, you position your organization not only to deliver measurable results but also to strengthen relationships and outpace competitors. Embracing this comprehensive, insight-driven framework is not merely a suggestion; it is a critical prerequisite for sustainable success in an increasingly dynamic and customer-centric business landscape.

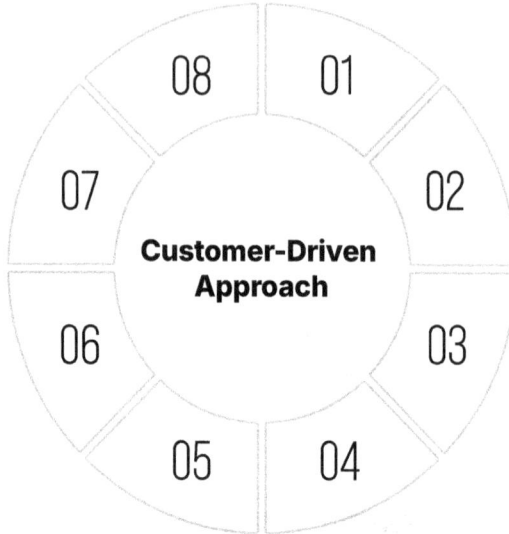

The Foundation for a Modernized Account Plan

01 Actionable Intelligence
Automated, actionable account intelligence mapped to relevant areas of focus

02 Prioritize Needs
Daily prioritization of relevant, customer-driven needs to create growth opportunities

03 Buyer Network
Buyer agreement network maps annotated with buying behavior dynamics

04 VIP Deep Dives
Deep stakeholder dives into important perceptions, backgrounds, and connections

05 Value Alignment
Alignment of capabilities and value to each stakeholder

06 Execution Progress
Daily monitoring of execution progress for each growth opportunity

07 Collaborative Workspaces
A collaborative space that unifies the team with up-to-the-moment, drill-down capabilities and tailored views

08 Customer Monitoring
AI-driven customer insights that monitor changing business and stakeholder needs and perspectives

7.4: Foundation for Modernized Plan

1. Automated Actionable Account Intelligence

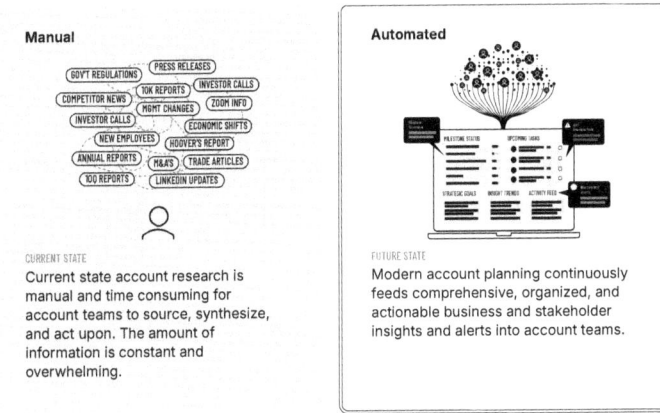

Manual

GOV'T REGULATIONS · PRESS RELEASES
COMPETITOR NEWS · 10K REPORTS · INVESTOR CALLS
MGMT CHANGES · ZOOM INFO
INVESTOR CALLS · ECONOMIC SHIFTS
NEW EMPLOYEES · HOOVER'S REPORT
ANNUAL REPORTS · H&A'S · TRADE ARTICLES
10Q REPORTS · LINKEDIN UPDATES

CURRENT STATE
Current state account research is manual and time consuming for account teams to source, synthesize, and act upon. The amount of information is constant and overwhelming.

Automated

MILESTONE STATUS   UPCOMING 1460
STRATEGIC GOALS   INSIGHT TRENDS   ACTIVITY FEED

FUTURE STATE
Modern account planning continuously feeds comprehensive, organized, and actionable business and stakeholder insights and alerts into account teams.

7.5: Automated Actionable Account Intel

Benefits of Automated, Actionable Intelligence to Account Teams

Teams can significantly enhance productivity by leveraging automated, targeted research delivery, eliminating the need for manual information searches. This approach offers two critical advantages:

- **Time Efficiency:** Automation streamlines data collection and analysis processes, freeing account teams to concentrate on strategic initiatives and deepen customer relationships.
- **Comprehensive Insight:** Continuous, automated monitoring captures evolving market trends, business dynamics, and actionable intelligence, equipping key account teams with a 360-degree perspective essential for driving account growth and retention.

> *"83 percent of sales leaders report their sellers struggle to adapt to changing customer needs and expectations."[19]*

By integrating automated research systems, organizations empower their teams to work smarter, respond faster, and make more informed decisions, transforming information overload into a strategic asset.

2. Daily Prioritization of Relevant Customer-Driven Needs

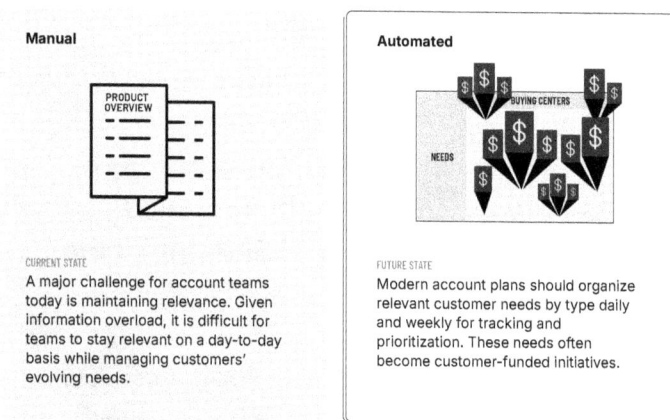

Manual

PRODUCT OVERVIEW

Automated

BUYING CENTERS

NEEDS

CURRENT STATE
A major challenge for account teams today is maintaining relevance. Given information overload, it is difficult for teams to stay relevant on a day-to-day basis while managing customers' evolving needs.

FUTURE STATE
Modern account plans should organize relevant customer needs by type daily and weekly for tracking and prioritization. These needs often become customer-funded initiatives.

7.6: Daily Prioritization of Relevant Customer-Driven Needs

Benefits of Daily Prioritization of Relevant Customer-Driven Needs

Teams can leverage newly found time savings to spend more time on new growth opportunities that are prioritized for activation.

---

19 Gartner, "Sales Performance: Top Leaders' Strategies," *Gartner*, accessed October 2, 2025, https://www.gartner.com/en/sales/topics/sales-performance.

- **Automate White Space Analysis:** Prioritization enables teams to organize and select the best areas for growth based on the buying centers and solution categories in which they fall. These prioritized opportunities can be reviewed and mapped into white space analyses and reports.
- **Increase Account Pipelines:** As opportunities evolve, account teams can directly feed these into account pipelines to increase quantity and quality.
- **Accelerate Speed to Need:** Based on specific market conditions and insights, address these challenges with customers before competitors do and solidify a strong, consultative role.

> *Only 28 percent of account management channels consistently achieve their cross-selling and account growth goals.*[20]

Note: Consider the role of a Commercial Insights Strategist. Upskilling SDRs, BDRs, researchers, and marketers with need-driven analysis and prioritization skills to feed actionable growth insights into account team sellers is an effective way of improving the workflow for optimal results.

---

20 Gartner, "Drive Growth Through Smarter Account Management," *Gartner*, accessed October 2, 2025, https://www.gartner.com/en/sales/trends/drive-growth-through-account-management.

3. Buyer Agreement Network Maps

Manual

Automated

CURRENT STATE
Org charts don't inform account teams about who is involved in making buying decisions or how all the stakeholders involved in a purchase decision interact and operate.

FUTURE STATE
Modern account plans include detailed maps of agreement networks across different buying centers that annotate the buying roles, interrelationships, and individual outcomes important to each stakeholder.

7.7: Buyer Agreement Network Maps

Benefits of Buyer Agreement Network Maps

Teams can focus on the complete view of the entire network of decision-makers involved in a deal, gaining valuable insights into who is part of the process, the decision-making roles they play, and how they interact with others in the network.

- **Increase Speed to Close:** By understanding the decision-making process and all those involved, account teams can expedite the process of reaching a decision.
- **Capture Valuable Tribal Knowledge:** The knowledge of who makes decisions and how they occur, once captured in key buying centers throughout the account, can be stored and maintained over time. This information can drive repeatable and effective growth initiatives with account stakeholders, positioning the company as a knowledgeable partner.

- **Reduce Onboarding Time:** As resources come in and out of the account, it's easier for teammates to get up to speed quickly and engage with customers from day one.

> *The average enterprise B2B buying group consists of five to eleven stakeholders, who represent an average of five distinct business functions. Meeting needs and driving consensus can be a challenge.*[21]

Note: You may consider adding a deeper-level view of customers' needs based on what's transpiring in individual departments regarding changing goals, expectations, and requirements. Constant change is happening more frequently today and is causing massive friction in the buying cycle.

4. Deep Dives into Important Stakeholders

Manual

Automated

CURRENT STATE
Customer relationships are often too limited or narrow, necessitating the expansion of these relationships to higher levels and across more buying centers, departments, regions, and subsidiaries.

FUTURE STATE
Modern account plans should provide deep dives on key stakeholders and reveal connections, relationships, and perspectives to support expansion efforts.

7.8: Deep Dives into Important Stakeholders

---

21 Gartner, "The B2B Buying Journey: Key Stages and How to Optimize Them," Gartner, accessed October 2, 2025, https://www.gartner.com/en/sales/insights/b2b-buying-journey.

Benefits of Deep Dives into Important Stakeholders

Teams can gain much greater insight into the key roles of the wallet owner, hero, guide, and other important influencers by increasing their level of understanding of the most important decision-makers.

- **Increase Stakeholder Knowledge:** By understanding the decision-makers involved at a detailed level, it's far easier to be relevant to their needs and valuable in providing insights that are meaningful to them, thereby building trusted relationships.
- **Leverage Connections:** Identify others involved in the account or industry to which they are connected and leverage these connections to more effectively influence and persuade key stakeholders.
- **Increase Deal Size and Close Rates:** When you are trusted by the decision-makers that matter, you will win more often and be awarded a bigger role in solving their challenges.

> *Only 29 percent of B2B buyers trust salespeople—but over 90 percent of buyers trust their peers.*[22]

---

22 Ian Bruce, "Who Do B2B Buyers Trust?," Forrester (blog), May 11, 2023, https://www.forrester.com/blogs/who-do-b2b-buyers-trust/.

5. Alignment of Capabilities, Resources, and Value

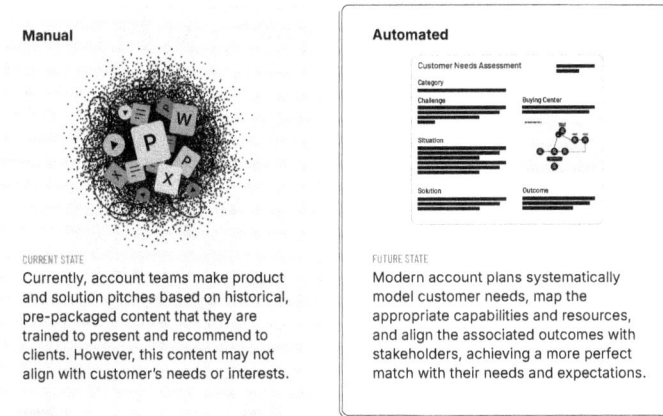

Manual

CURRENT STATE
Currently, account teams make product
and solution pitches based on historical,
pre-packaged content that they are
trained to present and recommend to
clients. However, this content may not
align with customer's needs or interests.

Automated

Customer Needs Assessment

Category

Challenge   Buying Center

Situation

Solution   Outcome

FUTURE STATE
Modern account plans systematically
model customer needs, map the
appropriate capabilities and resources,
and align the associated outcomes with
stakeholders, achieving a more perfect
match with their needs and expectations.

7.9: Alignment of Capabilities, Resources, and Value

Benefits of Alignment of Capabilities, Resources, and Value
Teams can achieve much greater alignment with customers by
systematically connecting solutions to their expressed needs in a
compelling manner and leveraging AI to summarize value within
the context of the situation.

- **Increase Deal Quality:** Aligning relevant capabilities,
  intellectual property (IP) and know-how, to customer
  needs with a compelling and sustainable value proposition
  improves the likelihood of closing deals, as the opportunity
  originates from the customer.
- **Reduce Meeting Prep and Follow-Up:** There is so much
  time that goes into developing meeting content, prepping
  the team, having the meetings, following up from the
  meetings, and adjusting as more information is provided.
  Tighter alignment up front reduces the time involved,

brings the right people into the customer conversation sooner, and drives a more productive conversation that reduces the follow-up time, as well as lost opportunity cost from being off message.

> *48 percent increase likelihood to grow an account when sellers focus on improving the customer's business.*[23]

6. Daily Monitoring of Execution Progress

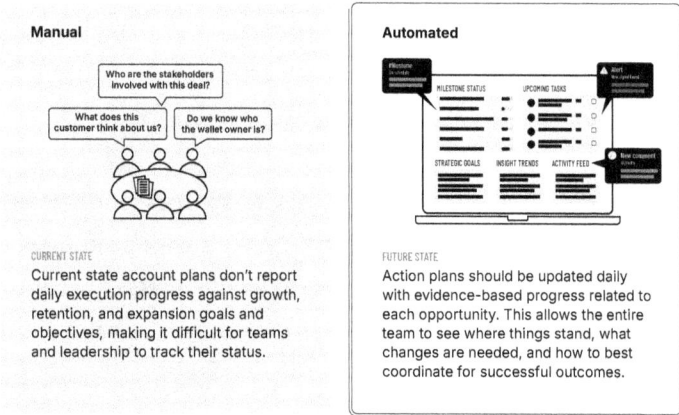

Manual	Automated
Who are the stakeholders involved with this deal? What does this customer think about us? Do we know who the wallet owner is?	Milestone status, Upcoming tasks, Alert, New comment, Strategic goals, Insight trends, Activity feed
CURRENT STATE	FUTURE STATE
Current state account plans don't report daily execution progress against growth, retention, and expansion goals and objectives, making it difficult for teams and leadership to track their status.	Action plans should be updated daily with evidence-based progress related to each opportunity. This allows the entire team to see where things stand, what changes are needed, and how to best coordinate for successful outcomes.

7.10: Daily Monitoring of Execution Progress

Benefits of Daily Monitoring of Execution Progress

Teams can align daily through online updates and drill-down views of key details regarding a particular pursuit and its current status. At the heart of an account plan is the progress made in new growth areas that are essential for expanding the relationship.

---

23 Gartner, "Customer Confidence," *Gartner*, accessed October 2, 2025, https://www.gartner.com/en/sales/insights/customer-confidence.

- **Increase Deal Velocity:** Increase the speed of execution through better overall orchestration of next steps with each customer, effectively managing all aspects of the decision-making process.
- **Improve Productivity:** Online dynamic account plans provide extensive productivity gains by reducing the time it takes to update each other, understand the current situation, and ensure everyone is aligned with the same set of facts.
- **Increase Coordination:** Teams are able to drive coordinated efforts more easily across different departments and functional silos.

> *"When account plans are frequently updated and account teams rely on them as a tool to drive decision making with the customer, organizations are 3x more likely to build customer decision confidence."[24]*

---

24 Gartner, "Account Management Strategies Every CSO Should Know," Gartner, accessed October 2, 2025, https://www.gartner.com/en/sales/topics/account-management-and-growth.

7. Collaborative Workspaces

Manual

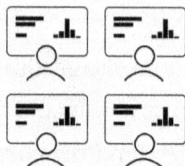

Automated

CURRENT STATE
Current state account plans do not provide a way for teams to collaborate on customer problems. Typically, these plans are maintained by the account seller as a tool to update leadership on account relationshipts rather than to manage the relationship itself with the account team.

FUTURE STATE
Modern account plans include collaborative features and support areas for joint problem-solving, sharing insights with customers, and storing account and deal assets to align and orchestrate value delivery.

7.11: Collaborative Workspaces

Benefits of a Collaborative Workspace

Teams can collaborate much more easily, which is particularly critical in large-scale customer accounts that span many silos of operations and internal resources.

- **Improve Innovation:** Better team collaboration increases innovation and creativity in solving customer challenges effectively. This results in better alignment of solutions and a clearer road map to valuable outcomes.
- **Increase Revenue:** Teams that collaborate internally and with customers can increase spending through a combination of a higher rate of engagement and greater confidence in the quality of engagement.

> *"We found that high levels of cross-functional collaboration increase the amount of key customer spend by 215 percent more than poor to mediocre levels of collaboration."* [25]

8. Effectively Using AI to Unlock Account Intelligence

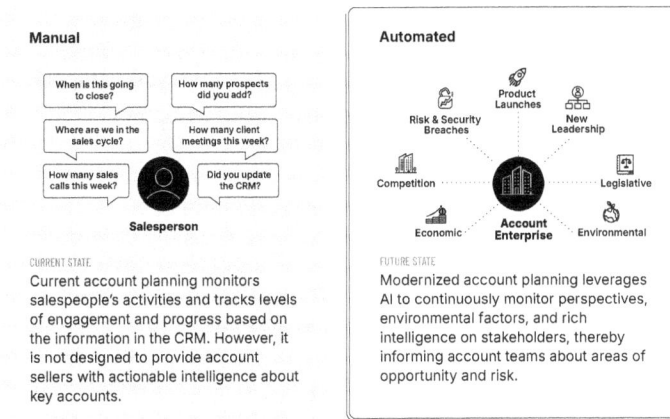

Manual

When is this going to close?

How many prospects did you add?

Where are we in the sales cycle?

How many client meetings this week?

How many sales calls this week?

Did you update the CRM?

Salesperson

CURRENT STATE
Current account planning monitors salespeople's activities and tracks levels of engagement and progress based on the information in the CRM. However, it is not designed to provide account sellers with actionable intelligence about key accounts.

Automated

Product Launches

Risk & Security Breaches

New Leadership

Competition

Legislative

Economic

Account Enterprise

Environmental

FUTURE STATE
Modernized account planning leverages AI to continuously monitor perspectives, environmental factors, and rich intelligence on stakeholders, thereby informing account teams about areas of opportunity and risk.

7.12: AI to Unlock Account Intelligence

Benefits of Using AI to Unlock Account Intelligence

AI is used to evaluate salesperson behavior based on CRM data and other customer interactions available through meetings and emails. Teams can unlock high-volume pipelines and engage much faster in the context of changing account conditions by using AI-enabled searching, summarization, and message alignment. These tools are tied to curated datasets associated with a broad view of data related to the account.

---

25 Gartner, "(Re)building Key Account Programs for Growth — Presentation Materials," *Gartner*, accessed October 2, 2025, https://www.gartner.com/en/documents/4007210.

- **Relevant Opportunity Identification:** Finding valuable insights is easily enabled when the right data sources are available to search in the first place.
- **Fast Insight Summarization:** Resources can simply ask AI chatbots to summarize the points of an article in seconds.
- **Tighter Value Alignment:** Connecting the key features of a broad set of capabilities to a specific customer need is simply summarized as an answer to a question.
- **Email Outreach:** Creating automated email outreach for a specific new opportunity is a simple request that accommodates preferences for style, length, and tone.
- **Revealing New Connections:** Asking AI to summarize connections between businesses, people, events, and any other set of objects is a quick query when the right dataset exists to search for these connections.
- **Best-Practice Recommendations:** When all the data is captured from account teams in online plans that track daily engagement efforts leading to successful outcomes, AI can make informed commercial recommendations based on similar situations and what has worked in the past.

DRIVING ACCOUNT GROWTH: FROM STRATEGY TO EXECUTION

Many organizations struggle to achieve consistent success in account planning and growth, often falling short of both business objectives and customer expectations. Recent Gartner research reveals a striking trend: 79 percent of sales organizations have overhauled their key account programs at least once in the past seven years due to persistent underperformance. This statistic is a clear warning—traditional approaches are no longer sufficient in today's dynamic market.

Modern account planning frameworks now offer advanced capabilities that accelerate growth, deepen client relationships, and proactively manage risk. The impact of adopting these new approaches can be both significant and immediate. Leading consultants such as McKinsey have identified seven best practices for the future of strategic account management. However, the real challenge and opportunity lie in moving beyond theory. Success depends on translating these insights into a dynamic, actionable account planning system that empowers teams to execute growth strategies consistently and effectively.

As you embark on this journey, remember: Sustainable growth is not achieved through sporadic initiatives or static plans. It requires a disciplined, adaptive process—one that continuously aligns with evolving customer needs and market realities. By embracing best practices and fostering a culture of execution, your organization can unlock new levels of performance, resilience, and client loyalty. Now is the time to transform account planning from a routine exercise into a powerful engine for growth.

Transform Account Planning

- ☑ Use analytics to mine cross-sell opportunities
- ☑ Manage stakeholder complexity
- ☑ Understand the quality of stakeholder relationships
- ☑ Mine more data sources for insights and action
- ☑ Use digital account planning to sync the team
- ☑ Empower teams to innovate
- ☑ Be releevant to buyers and very personalized

7.13: Transform Account Planning Checklist

ROLE BY ROLE: STEPS TO SHAPE THE FUTURE

For Account Teams

Modernized account plans free up a tremendous amount of time. With a more insight-driven, dynamic planning and execution team experience, you will be able to increase the volume and velocity of deal pursuits, adapt much more quickly in the moment, and stay relevant with customers at all times. It will be important to lead and inform your team more frequently, with a faster cadence, but fewer meetings and more digital updates on progress, priorities, and results achieved. The ability to get in the flow, aligned with the rhythms of your account relationships, is an opportunity to dramatically change the dynamic and increase the energy, momentum, and overall health of your account relationships.

For Marketing

Marketing has historically not participated in account planning. However, with the increasing speed of customer interactions and deeper insights into account needs, marketing can now tap into this new source of enterprise account interests, behaviors, successes, behavioral dynamics, and areas of traction or lack thereof. This information can be used to better inform marketing strategies, communications, and programs.

For Management

Management can continuously monitor the status of each customer to make faster, better-informed decisions about investing human capital and focus. This helps optimize revenue growth, reduce risks, and protect your greatest asset—your customers. You'll be able to see customer complaints by stakeholder much more quickly, enabling the organization to implement necessary policy changes that remove obstacles for your teams. This also supports investing in new capabilities and encourages the development of playbooks that better align with customer needs.

For Sales Enablement

Account planning has historically been a massive time sink and a deeply resented, manual process. Sales enablement should focus on revenue enablement across both sales and account teams. This is where you can help drive an additive pipeline that is high in both value and quality.

Balance coverage between sales teams focused on prospecting and those focused on account growth by implementing capabilities that enable online account planning, making it easier, less burdensome, and less time-consuming. Integrate data into account plans to inform sales teams about what to pursue and where the best opportunities lie, instead of expecting them to figure everything out on their own.

For Commercial Insight Strategists

Your role focuses on growth opportunity prioritization. With Bluespace serving as the backbone for the team's entire action plan, you drive the prioritization of the team's focus areas. Your opportunity is to act as the intelligence arm of the account team—providing strategic insights to achieve account growth goals and taking a leadership position in doing so. Your career path is to become an account executive by becoming an expert on the customer.

Before we dive into the next chapter, pause and ask yourself: What if the real breakthrough in account growth isn't about the tools or the strategies, but about truly seeing the world through your customer's eyes?

You've seen why traditional approaches no longer cut it and how best practices set the stage for success. But theory alone won't move the needle. The real transformation begins when you step into the trenches—when the rubber meets the road, and you're faced with messy, real-world challenges like relentless requests for proposals (RFPs), skeptical clients, and the pressure to deliver outcomes that matter.

Ready to discover what happens when you shift from planning to action and what it takes to stand out in a sea of sameness? Let's turn the page and see how one pivotal conversation with a frustrated customer changed everything. The journey from routine to remarkable starts now.

CHAPTER 8:

WHAT CUSTOMERS VALUE

In the chapters ahead, we will explore a critical shift in how organizations must align their capabilities, resources, and value with what customers truly need and expect. This is no longer a static exercise. It's a dynamic, ongoing process of adaptive alignment. As customer environments, priorities, and stakeholders evolve, so must our ability to stay in sync. In today's competitive landscape, the companies that modernize their systems, processes, and skill sets to operate at the speed of their customers will not only grow faster—they'll outpace and displace competitors who fail to keep up.

I learned this lesson the hard way. Another RFP landed on my desk with a tight deadline, little insight into the customer's real needs, and even less confidence in winning. Still, we responded, just like we always did. This one came from a major retailer investing in a new omnichannel marketing platform. We followed the standard playbook: polished technical documentation, detailed implementation plans, and a showcase of our team's capabilities. It was thorough. It was professional. It was exactly what everyone else was doing.

Then came the moment that changed everything. During one of the review meetings, the customer, clearly exasperated, said something I'll never forget: "All you guys are the same. We don't care about the technology—we can't even tell the difference. What we want is someone who can run the campaigns and get the results. Can you do that?"

It was a wake-up call. For years we had been leading with technology, not outcomes. We were selling the engine, not the results. That comment forced me to confront a painful truth: We weren't aligning with what the customer valued. And what they valued wasn't the technology. It was the outcomes driven by the technology. We had assumed they knew what to do with our solution, but they really didn't. They were retail experts, not campaign experts.

From that moment on, we flipped our approach. We led with campaign performance, not product specs. We showcased the processes, people, and analytics that would drive measurable results. We stopped talking about what our technology could do and started proving what we would deliver. We used our own tools to run real campaigns, analyze performance, and continuously improve results.

The transformation was fast and dramatic. We went from being seen as a reliable but minor player to a serious competitor. Our win rate jumped tenfold. We started beating larger firms still stuck in the old model. Analyst reviews improved. Customers noticed. And all because we stopped assuming what they wanted and started listening to what they actually needed.

This experience taught me that real differentiation doesn't come from what you build. It comes from how well you align with what your customers are trying to achieve. If you want to win in today's market, don't just sell what you have. Solve what they value.

REFRAMING THE PRODUCT MINDSET: A STRATEGIC IMPERATIVE

In today's market, the tendency to overvalue our own products, what we call "productitis," is a common but costly misstep. It stems from a deeply ingrained mindset: the belief that our offerings are inherently exceptional and that customers will naturally recognize their value. The reality is far different.

Customers aren't buying products. They're seeking solutions. Regardless of how innovative or feature rich your product may be, most buyers can't distinguish it from the competition. They're unlikely to invest the time or energy to do so. The marketplace is saturated with similar offerings, and to the customer it often feels like choosing between twenty indistinguishable options on a grocery store shelf.

This sameness has made traditional product-centric selling obsolete. Buyers have evolved. They conduct their own research, form opinions independently, and increasingly avoid sales conversations that revolve around product features.

The Shift: From Product to Purpose

To succeed in this environment, we must shift our focus from what we sell to why it matters. This means aligning with customer priorities, understanding their business challenges, and positioning our offerings as enablers of their success, not just as products, but as strategic solutions.

This isn't just a sales tactic—it's a mindset. It requires curiosity, empathy, and a commitment to delivering value beyond the product itself. When we lead with insight, not inventory, we earn trust. When we speak about outcomes, not options, we gain relevance.

A Call to Action

Let this be your call to break free from the product-first mentality. Challenge assumptions. Ask better questions. Seek to understand before you seek to sell. The future belongs to those who can connect with customers on a deeper level—not through features, but through foresight.

The opportunity is clear: Move from product obsession to customer obsession. That's how we differentiate. That's how we win.

ADDRESSING CUSTOMER MISALIGNMENT: RECOGNIZING THE ROOT CAUSES

In the early days of industry, business was straightforward: Companies designed products, developed go-to-market strategies, and assigned quotas by territory. Customer needs were simpler, and product-centric approaches sufficed.

Today the landscape is unrecognizable by comparison. Customer challenges are multifaceted, organizations are sprawling across business units and geographies, and digital commerce has introduced unprecedented complexity. Yet many organizations persist with outdated go-to-market mindsets, failing to adapt to the realities of modern markets.

Only 28 percent of account management channels consistently achieve their cross-selling and growth goals.[26]

The result is a widening gap between what suppliers offer and what customers actually need—a gap that continues to grow in the absence of meaningful change.

The Modern Enterprise: Complexity and Silos

Organizational silos are no longer confined to departments; they now span newly acquired business units, each with its own priorities and processes. Marketing and product teams often operate in isolation, focusing on mergers and acquisitions to drive revenue rather than fostering organic growth. Sales teams are left to "figure it out," tasked with cross-selling into "white space" without a unified view of customer needs. Too often only account sellers, those closest to the customer, have any real understanding of what buyers want, and even then, their insights rarely reach senior leadership.

The Persistence of Outdated Approaches

Despite these challenges, management frequently clings to legacy strategies, reinforced by habit rather than insight. Customer feedback mechanisms remain product-centric, asking, "What do you think of our product?" instead of the more critical, "What needs do you have that we can help address?" Leadership may overlook the true source of underperformance, attributing missed targets to sales execution rather than systemic misalignment with customer

---

26 Gartner, "Drive Growth Through Smarter Account Management," *Gartner*, accessed October 2, 2025, https://www.gartner.com/en/sales/trends/drive-growth-through-account-management.

needs. Account sellers become mere messengers, pitching what the company provides, not what the customer demands.

A Call to Action: Shifting Mindsets and Embracing Opportunity

This is a pivotal moment. With the advent of AI and advanced analytics, leaders have a unique opportunity to realign solutions with customer needs and empower account sellers with new, effective workflows. However, realizing this potential requires a fundamental shift in mindset—one that prioritizes serving customers above meeting internal sales targets.

Leaders must begin with the customer and work backward, designing strategies that solve real problems rather than pushing products to meet financial forecasts. This approach demands courage, discipline, and a willingness to challenge established norms.

The Cost of Misalignment

Misalignment is not a minor issue; it is expansive and multidimensional. Recent research from Corporate Visions highlights that:

> *B2B buyers frequently stick with existing solutions or develop in-house alternatives when external vendors fail to demonstrate a clear path forward. Even when buyers express interest, deals often stall because proposed solutions do not fit evolving needs or because sellers cannot articulate tangible value. Too many sales reps focus on features rather than business fit and measurable outcomes, perpetuating the cycle of misalignment.[27]*

---

27 Anton Rius, "B2B Buying Behavior in 2025: 40 Stats and Five Hard Truths That Sales Can't Ignor," *Corporate Visions* (blog), January 14, 2025, https://corporatevisions.com/blog/b2b-buying-behavior-statistics-trends/.

Consider the lost opportunity cost: Imagine your win rates are only 30 percent of what they could be, along with your average deal size. What would that add up to? It would be astronomical—and this is corrected with alignment, not incremental investment.

THE CRITICAL ROLE OF SOLUTION ALIGNMENT

Success hinges on one decisive factor:

**Aligning your solution precisely
with the customer's needs.**

Its impact has only grown stronger, surpassing all other factors in determining outcomes.

When evaluating why deals are won or lost, the ability to fit your solution to the client's requirements emerges as the most significant predictor. This decisive factor is a call to action. Organizations that consistently ensure their offerings address the true needs of their customers position themselves to outperform competitors and secure lasting relationships.

Key Drivers of Wins and Losses

Many elements influence success, but Corporate Visions' published research shows executive buyers rank these decision-making factors as most critical, in order of importance:

- Align solution to needs.
- Articulate meaningful value.
- Negotiate creatively.
- Deliver compelling communications.
- Demonstrate clear differentiation.

- Help justify decisions.
- Provide expert insights.
- Resolve concerns responsively.

The research clearly concluded that the single biggest predictor of sales success is **solution alignment**—aligning your solution to buyer needs. Based on buyer feedback from 120,000 win/loss reports over a two-year period, this one competency has:

> - *"Emerged as the top driver of deal outcomes across wins, losses, and no-decisions—becoming more influential than any other sales competency.*
> - *Shown surprising independence from other skills. While most competencies work in concert, solution alignment increasingly drives decisions on its own.*
> - *Demonstrated consistent growth in importance with new data from each quarter, confirming it's not a temporary trend but rather a fundamental shift in buying behavior."*[28]

Getting the solution aligned is the cornerstone—without it, strengths elsewhere won't make up the gap.

Aligning Solutions with Customer Needs

Achieving alignment between solutions and customer needs is not only more effective but also less burdensome than maintaining misalignment. Removing organizational silos fosters efficiency and innovation, while collaboration across departments accelerates progress and unlocks new opportunities. Listening to customers

---

28 Corporate Visions, "When "Solution Fit" Fails, Nothing Else Works," Winsight research brief (n.p.: Corporate Visions, n.d.), PDF. https://corporatevisions.com/resources/sales-skills/when-solution-fit-fails-nothing-else-works/.

requires less effort than avoiding their feedback, and it yields far greater rewards.

The integration of artificial intelligence enables organizations to combine diverse capabilities into tailored solutions that address specific customer outcomes. By leveraging both historical and real-time customer input, automation can streamline these processes, liberating employees from restrictive workflows and empowering them to serve customers with creativity and agility.

Empowering Account Sellers

When account sellers are consulted, their requests consistently center on eliminating obstacles that hinder their performance. They require the following:

- Support systems that simplify complexity.
- Automation that delivers relevant solutions precisely when needed.
- Tools that facilitate clear communication and rapid delivery of customer-centric solutions.

Providing these resources enhances productivity and ensures that solutions are delivered at scale and speed, directly aligned with customer expectations.

The Path Forward

This approach embodies the core principles of a robust Go to Customer strategy and the Selling What's Possible™ methodology. By prioritizing alignment, collaboration, and customer-centric innovation, organizations position themselves to achieve sustainable growth and lasting customer loyalty. The choice is clear:

Empower teams to break free from outdated barriers and enable them to deliver what customers truly need.

MATCHING SOLUTIONS TO NEEDS: BEGIN WITH A CAPABILITIES INVENTORY

Understanding and aligning organizational capabilities with customer needs is essential for delivering value and building strong client relationships. While suppliers possess a deep knowledge of their own offerings, there is often a disconnect between what organizations value internally and what customers seek.

Overcoming Internal Bias

When asked to list company capabilities, organizations frequently generate a wide range of responses, many of which are overly detailed, biased by departmental perspectives, or focused on technical minutiae. This tendency, "productitis," can cloud judgment and obscure what truly matters to customers. It is important to recognize that one's position within the company shapes internal viewpoints, but these perspectives rarely mirror the priorities of decision-makers on the customer side.

Note:

- **Technically.** A capability is any form of intellectual property your company possesses. Since there are so many different forms of businesses, there are many different intellectual property laws and business models based on those patterns.
- **Simply.** A capability is an ingredient that can be combined with other ingredients to add more value based on a given customer's situation.

Adopting a Customer-Centric Perspective

To bridge this gap, organizations must adopt a top-down approach. The focus should shift from internal features to the attributes customers genuinely value:

- **Trust and Reputation:** Customers prioritize working with vendors who are recognized as trustworthy and competent in their field.
- **Proven Track Record:** Demonstrated success in delivering outcomes for similar clients is highly persuasive.
- **Comprehensive Capabilities:** Skills, processes, experience, and methodologies must be evident and aligned with client objectives.
- **Innovative Approach:** The ability to bring vision, new ideas, and robust solutions to the table sets partners apart.
- **Breadth and Depth:** A strong portfolio of features, functions, strategic partnerships, and data-driven insights enhances credibility.

Customers Value Capabilities That Span Many Areas

☑ Functionality	☑ Labor	☑ Research
☑ Knowledge	☑ Approaches	☑ Data
☑ Skills	☑ Case Studies	☑ Ideas
☑ Processes	☑ Expertise	☑ Access
☑ Features	☑ Vision	☑ Customers

8.1: Customers Value Capabilities That Span Many Areas of IP

UNLOCK YOUR TEAM'S KNOW-HOW WITH AI-DRIVEN CAPABILITY MAPPING

Building a dynamic, AI-accessible capability inventory is essential for organizations aiming to stay competitive and innovative. By thoughtfully organizing your products, services, and areas of expertise into a clear inventory, you enable a seamless combination of offerings to suit evolving customer needs. Take a critical look at your current capability list—don't limit yourself to existing structures. Consider how new services or solutions can be bundled with your technologies, enhancing customer value by relieving them of complex management responsibilities.

Capability Map

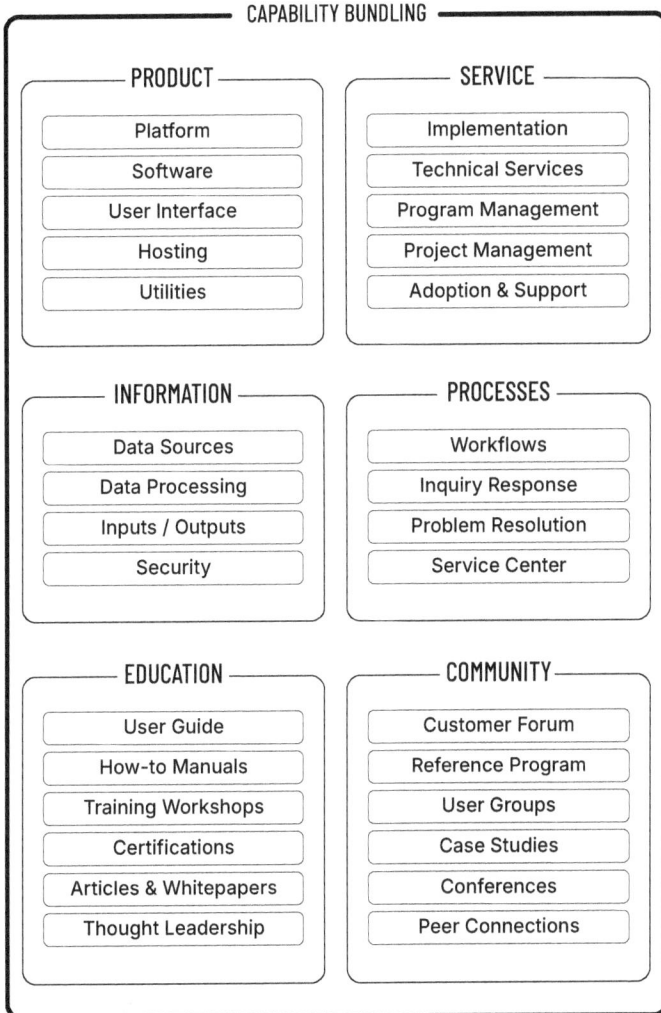

CAPABILITY BUNDLING

PRODUCT
- Platform
- Software
- User Interface
- Hosting
- Utilities

SERVICE
- Implementation
- Technical Services
- Program Management
- Project Management
- Adoption & Support

INFORMATION
- Data Sources
- Data Processing
- Inputs / Outputs
- Security

PROCESSES
- Workflows
- Inquiry Response
- Problem Resolution
- Service Center

EDUCATION
- User Guide
- How-to Manuals
- Training Workshops
- Certifications
- Articles & Whitepapers
- Thought Leadership

COMMUNITY
- Customer Forum
- Reference Program
- User Groups
- Case Studies
- Conferences
- Peer Connections

8.2: Example Capability Map

Now is the time to identify untapped opportunities. Providing expertise as specialized on-demand resources or retainer-based solutions not only differentiates your business but also opens up new streams of revenue. For instance, adding campaign management services to support your campaign technology proved highly successful, yielding incremental fees and delighted clients.

This work is not a one-time project; it's an ongoing strategic practice. Success requires close collaboration among marketing, finance, and product teams. Regularly gather customer feedback and analyze success stories to spark the creation of new, desirable capabilities with measurable outcomes. Make each thoughtfully developed capability readily accessible within your inventory to ensure your organization consistently delivers the unique value clients deserve. Start this process now—success hinges on agility, innovation, and relentless focus on customer impact.

Connect Capabilities to Outcomes

After establishing your list of capabilities—ideally organized within broad, intuitive categories to keep them accessible and actionable—the next essential step is to connect each capability to a clear, meaningful outcome. This alignment is not just a best practice; it's fundamental to delivering real value for your customers. By mapping outcomes to capabilities, you create a direct line between what you offer and what your customers genuinely need.

It's important to remember that you rarely understand a customer's true end goals right from the start. Preliminary signals only provide hints. Through intentional discovery, ongoing dialogue, and collaborative engagement, you uncover deeper insights into their challenges and ambitions. As these emerge, you're equipped to strategically combine the right capabilities in a way that not only addresses surface-level needs but delivers on the outcomes that matter most to them.

Approaching solution design from this perspective ensures you remain flexible, responsive, and laser-focused on customer success. The work may seem demanding, but each thoughtful connection between a capability and a desired outcome brings you closer to building trust, strengthening relationships, and producing measurable impact. Ultimately, this rigorous, customer-centered thinking is what turns potential into performance.

Creating a simple inventory might look something like this.

PRODUCT		SERVICES		EXPERTISE	
Product	**Outcome**	**Service**	**Outcome**	**Expertise**	**Outcome**
Product A	Outcome A	Service A	Outcome A	Expertise A	Outcome A
Product B	Outcome B	Service B	Outcome B	Expertise B	Outcome B
Product C	Outcome C	Service C	Outcome C	Expertise C	Outcome C

8.3: Capability and Outcome Inventory

WHAT IS AN OUTCOME?

An outcome, particularly from a customer's perspective, extends far beyond simple financial impacts. In practice, outcomes encompass a range of attributes, such as convenience, reliability, quality, and overall satisfaction, that executives carefully evaluate when determining the true value of a product or service. While this might initially sound theoretical, it's actually a fundamental part of our everyday decision-making process. We all instinctively assess value through multiple lenses, not just price.

Recognizing this, it becomes essential to consider the full spectrum of customer outcomes, because understanding and delivering on what truly matters to customers is what sets exceptional organizations apart. By broadening your perspective and appreciating the diverse factors that contribute to customer value, you unlock the potential to drive meaningful impact and lasting success.

There are three correlated areas:

1. What is the future state and is the value of achieving the result of the future state worth the effort?
2. How will this future state be achieved and when will we achieve it?
3. Who is the executive owner and where are the impacted stakeholders in the company?

The Outcome Wheel

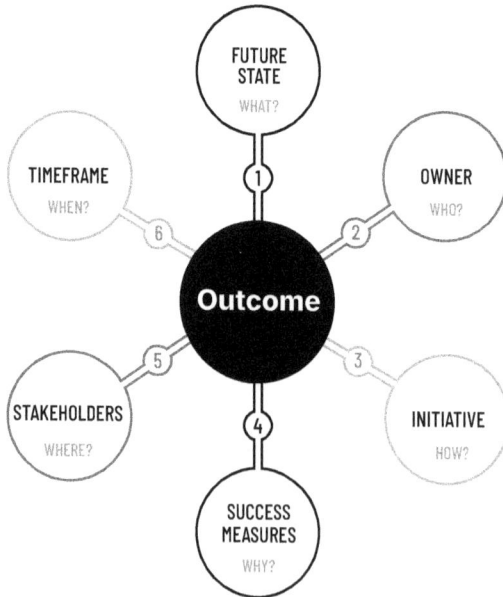

8.4: The Definition of an Outcome

Critical Considerations for Delivering Value Aligned to Customer Outcomes

Highlighting return on investment (ROI) without anchoring it to a clear time frame or budget cycle can undermine executive confidence and decision-making. Executives require a comprehensive framework that not only quantifies value but also contextualizes how and when that value will be realized. To promote clarity and informed decision-making, outcomes should be defined with explicit attributes—before associating them with specific products, services, or expertise.

Example: Structuring Outcome Attributes for a Service

- **Service Name:** Pipeline Opportunity Prioritization
- **Desired State:** Prioritized Opportunities Delivered as a Service
- **Accountable Owner:** CRO
- **Strategic Initiative:** Enhance Pipeline Quality
- **Success Metric:** Achieve 80 Percent Close Rate
- **Key Stakeholders:** Account Sellers
- **Time Frame:** Three Months

Leveraging AI-Driven Bundling for Customer Outcomes

Building an inventory that links products, services, and expertise to clearly defined outcome attributes enables organizations to lead with customer outcomes and intelligently bundle capabilities, especially when integrating generative AI. This approach aligns solution-selling with what customers value most and positions offerings for maximum impact.

The Power of Prepackaged Bundles

Account sellers often show strong enthusiasm for leveraging prepackaged solution bundles that have a proven track record of successful sales and delivery. These pre-bundles serve as an elevated tier of capabilities, much like choosing a ready-made dinner from the grocery store instead of selecting individual ingredients. This approach is critically important because these bundles are strategically engineered to align with common customer needs and desired outcomes, ensuring a more seamless and effective sales process.

However, a common challenge arises as many sellers remain focused on selling smaller, individual product components. This focus can inadvertently limit their engagement to lower organizational levels within the customer's hierarchy, often without their awareness. This phenomenon relates directly to the "gorilla" effect discussed in chapter 6: the tendency to miss or ignore opportunities at higher levels of urgency and strategic importance.

To maximize impact and revenue potential, it is essential to equip account sellers with pre-bundles that correspond to higher-altitude customer needs. When the customer's urgency and complexity escalate, providing a comprehensive pre-bundle—integrating products, services, and expertise that directly address these elevated outcomes—empowers sellers to align with the buyer's priorities and engage meaningfully at the appropriate organizational level. This strategic alignment not only drives stronger business outcomes but also advances the seller's positioning as a trusted adviser capable of delivering impactful solutions.

Establishing Repeatable Patterns Through Outcome Mapping

A reverse-engineered approach, starting with the outcomes customers seek, enables organizations to identify and replicate winning solutions by answering key questions:

- Who are the key stakeholders seeking these outcomes?
- Which capabilities have consistently delivered such outcomes?
- What success rates, margins, pricing, costs, and time-to-results have comparable solutions achieved?
- Is there documented evidence of proven customer impact and successful delivery?

Proactively mapping buying centers, altitude levels, and outcome patterns creates a blueprint for scalable, repeatable success. It empowers account executives with tested recommendation sets, increasing both confidence and effectiveness in engaging executive stakeholders.

Mapping the Buying Center

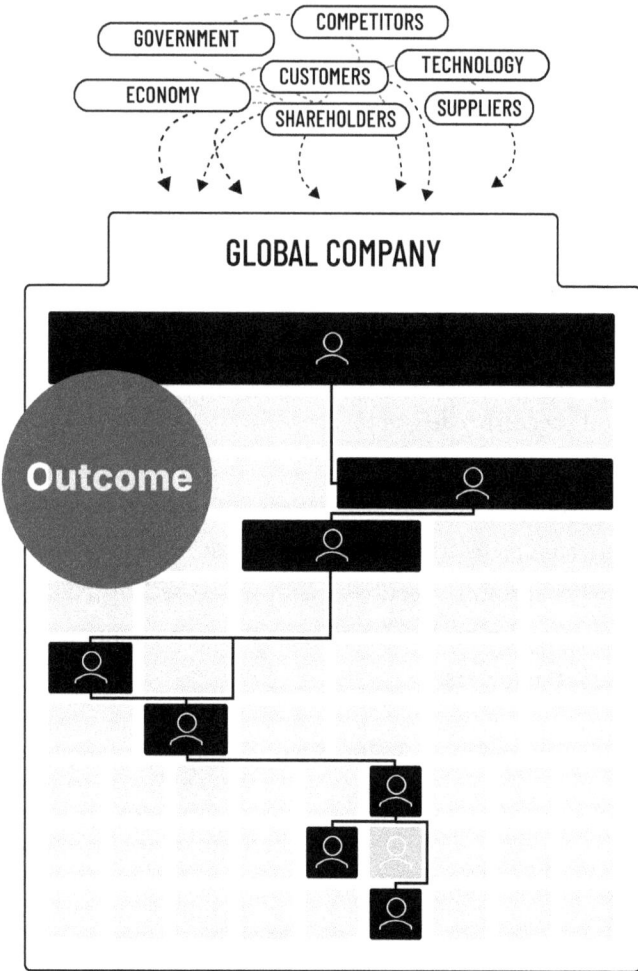

8.5: Mapping Decision Makers in a Buying Center

OPTIMIZING BUNDLE PERFORMANCE: A STRATEGIC IMPERATIVE FOR ENTERPRISE GROWTH

To drive sustainable growth and competitive advantage, companies must rigorously track and analyze the performance of capability bundles proposed to customers. Win-loss analyses that focus on bundles (comprehensive sets of capabilities) offer a streamlined, insightful approach to understanding which combinations deliver the greatest business outcomes.

Why Tracking Bundles Matter

Tracking how often specific bundles are proposed, purchased, and successfully delivered, alongside the outcomes realized, provides critical data that fuels AI-driven sales optimization. These datasets empower advanced algorithms to recommend precisely tailored bundles matching evolving customer needs, thereby maximizing the likelihood of closing deals. This approach represents the new frontier of insight-driven selling in large enterprise accounts, where complexity demands precision.

Key Principles for Leveraging Bundles Effectively

1. **Expertise Over Product:** Customers increasingly value the expertise guiding them through complex solutions as much as the products themselves. Position knowledge and advisory capabilities at the core of the sales journey to deliver meaningful impact.

2. **Innovation as a Constant:** Innovation remains paramount. Gartner consistently highlights the rapid evolution of customer expectations fueled by technology—especially AI—and shifting behaviors. Continuously refine existing

capabilities, create new bundles, and offer choices that align with these dynamic requirements to stay ahead.

3. **Data Integrity Fuels AI Effectiveness:** Maintaining up-to-date, accurate data on capabilities and their delivered outcomes is essential. Clean datasets enable AI to make reliable, context-sensitive recommendations that resonate with client needs in real time.

4. **Bundling Enhances Margins:** As transactional sales face margin compression and automation, capability bundles become vital to preserving and enhancing profitability. Bundles that deliver superior customer outcomes justify premium pricing, thus driving higher margins and shareholder value.

5. **Accelerated Time to Value:** Buyers increasingly prioritize asset-based services—leveraging intellectual property and expert know-how—that deliver faster, measurable business results. Bundled solutions that solve complex problems efficiently position providers as indispensable partners.

6. **Leverage Historical Successes:** Using data from past wins to inform future engagements maximizes success rates. Aligning bundles to customer context and proven outcomes ensures relevance and boosts confidence in proposed solutions.

ROLE BY ROLE: STEPS TO SHAPE THE FUTURE

For Account Teams

Imagine a future where intelligent systems seamlessly match the ideal capabilities to your customers' precise needs. To thrive in this landscape, it is essential to be meticulous and deliberate in capturing customer requirements with accuracy and depth. Customers communicate their needs most effectively through the outcomes they aspire to achieve. Your role is to convincingly sell the gap, the difference between their current situation and their desired future state, while emphasizing the urgency and risks associated with leaving that gap unaddressed. Success depends on fully understanding all stakeholders involved and carefully assessing their unique objectives. By doing so, you dramatically increase your likelihood of winning and sustaining business over the long term.

For Marketing

Marketing must evolve beyond traditional approaches by partnering closely with product teams and finance leaders to design compelling capability bundles. The most valuable customer insights often come from your largest clients, who act as bellwethers for the broader industry. Use these insights as the foundation to innovate, crafting new combinations of products, services, and expertise that directly respond to market needs. Your challenge is to create packages and pricing models that clearly demonstrate the value linked to the outcomes delivered, making it easier for customers to understand and justify their investment.

For Management

Breaking down internal silos is not merely advisable; it is mission critical. Obstacles that reinforce isolation between departments undermine collaboration, stifle innovation, and ultimately weaken performance. Companies that thrive unify their teams around a shared principle: working backward from the customer. This approach mandates cross-departmental cooperation, fostering engagement and collective ownership of corporate goals. To embed this culture, consider redefining objectives, goals, and incentive structures so they reinforce shared accountability. When teams succeed together, the business succeeds overwhelmingly.

For Sales Enablement

Sales enablement professionals are the champions for this transformative vision. Your role is to recommend robust frameworks, select the right tools, and identify strategic partners who can help design and implement new processes that align capabilities with customer needs. Crucially, the solutions you facilitate must be intuitive and actionable for account sellers in the field. Focus intently on how success will be measured, not just in outputs, but through tangible impact on sales performance and customer satisfaction. Your influence will determine how effectively this vision is executed and embraced.

For Commercial Insight Strategists

As the architects of alignment, you bear the vital responsibility of ensuring every opportunity is tightly calibrated between customer needs and recommended solutions. This means rigorously identifying gaps in understanding and thoroughly validating customer outcomes. Are all stakeholder perspectives incorporated into the solution? Your insight leadership is indispensable to guiding account executives to consistently deliver actionable, precise solution recommendations from the outset. By anchoring the commercial process in deep insight, you empower the entire team to win bigger, faster, and more frequently.

As we've seen, the intelligent use of data and AI to tailor capability bundles sets the stage for winning deals and driving growth, but there's an equally critical piece to the puzzle that often goes overlooked: the human element. Because no matter how sophisticated your analytics or how innovative your offerings, it's the people behind the capabilities who truly shape customer value and competitive differentiation. So, what happens when the very expertise your customers depend on is tangled in internal roadblocks or misaligned priorities? In the next chapter, we'll explore why "who" delivers your solutions matters as much as "what" you're selling and how organizations that get this right unlock not just sales, but lasting trust and strategic advantage.

CHAPTER 9:
WHO CUSTOMERS VALUE

When the account manager encountered the introduction of a new resource request process by internal leaders, their goal was to streamline support for account teams. Instead, resource managers created a bottleneck: Every request now required a formal submission, with a four-week turnaround just to confirm resource availability. This left the account teams caught between two difficult paths. They could either adhere to the slow process and risk disappointing customers, losing deals, and damaging their credibility, or they could avoid resource-dependent opportunities altogether and focus solely on low-margin, commodity sales. The latter strategy meant selling more product deals that were smaller and less valuable, just to meet quotas.

Meanwhile, more agile and customer-focused competitors stepped in to fill the strategic gaps. These competitors made expertise readily accessible, winning key opportunities and leaving the account teams at this provider as the bureaucratic vendor on the sidelines, often the last option in procurement discussions. Management's passive endorsement of this process only deepened the problem. They pointed fingers at the account teams, blaming them for poor results. This drove many top account leaders, to consider joining competitors, encouraged by frustrated customers.

This scenario serves as a cautionary tale of how rigid, internally driven processes can undermine business success, effectively calling in an airstrike on one's own position. As I mentioned earlier, today's buyers prioritize expertise over products. If an organization becomes the gatekeeper or, worse, the preventer of that expertise, it will inevitably degrade into a commodity-focused firm operating on thin margins. For public companies, this translates directly into declining shareholder value, stagnant revenues, and pressure to rely heavily on sales automation and AI to sustain profitability. The message is clear: Agility, customer responsiveness, and seamless access to expertise are indispensable for sustainable growth in today's competitive market.

WHO YOU ARE MATTERS AS MUCH AS WHAT YOU OFFER

Customers increasingly prioritize the expertise and knowledge of their vendors above the products themselves. Specialists, those with deep, multidisciplinary skills, are the true differentiators who drive high-value outcomes. The quality of services and solutions hinges on engaging the right people at every stage.

To seize strategic, higher-value deals that deliver measurable business improvements, organizations must ensure the right experts are involved and that their time is wisely invested. This approach is not about cost cutting but about investing in capabilities that create meaningful customer outcomes and long-term success.

One customer's perspective underscores this dynamic perfectly:

"Oh, absolutely, and it's even more than that. Once we trust a given vendor, we'll bounce ideas off them before talking to anyone else. If they can help us, they tell us, and we just work with them without even telling anyone else."

This level of trust and collaboration is the ideal position for any account team and leadership. Imagine a partnership where customers confide exclusively in you, free from competitive interference. This happens because vendors who fail to collaborate impose burdens on customers, who then choose easier-to-use alternatives. Conversely, when expert teams deliver seamless, collaborative support, financial results improve significantly, reflecting a strong return on expert engagement.

The tangible benefits are clear:

- Larger and more strategic deals.
- Stronger, sustainable revenue streams.
- Extended contract durations.
- Higher profit margins.
- Increased repeat business.
- Growing deal sizes.
- Satisfied, loyal customers.
- Trusted, referenceable partners.

Who wouldn't want these outcomes?

Consider a proven example where a diverse team of five specialists with expertise in technical, consulting, analytics, strategy, and commercial domains focused exclusively on one major account. Over five years, this team generated more than $150 million in contract value across seven divisions domestically, managing more than 250 statements of work under a master services agreement and coordinating with roughly two hundred stakeholders. The average deal size was around $250,000, with many larger contracts closing successfully.

In stark contrast, a larger group of sixteen salespeople conducting product-centric, transactional sales delivered approximately one-tenth of the revenue at ten times the cost. Their average deal size of this group was only $10,000, requiring twenty-five transactions to match just one enterprise account sale. Moreover, this transactional team faced relentless challenges from commoditization, price pressure, regulatory complexity, and competition—factors that make this sales model increasingly vulnerable to automation and AI displacement. In fact they had to constantly re-sell the same commodity order over and over against the competition which took up most of their time.

By focusing on blended expert teams addressing strategic enterprise needs, organizations achieve significantly better results at lower costs and with higher margins. This is the future of account growth: moving away from fragmented transaction-based selling toward a holistic, value-driven approach that views the entire account's potential strategically.

The challenge for management is clear: Reimagine sales as team-driven, expertise-led collaboration rather than an individual transaction chasing exercise. Ask critical questions:

- Who is on the account team?

- Are the right skills and domain knowledge represented?
- How well is the team aligned with the customer's broader growth agenda?

Breaking free from the transaction mindset to embrace strategic, expert engagement unlocks the full growth potential within enterprise accounts. This is imperative for sustaining success in today's market.

RECOGNIZING TRUE VALUE: BEYOND JOB TITLES AND HISTORIES

Many organizations mistakenly focus on candidates' job titles or employment history when evaluating potential value. Yet, from a customer's perspective, these factors rarely translate to meaningful impact, except perhaps in terms of formal decision-making authority. Ineffective performers and mismatched stakeholders often share impressive job histories and titles but fail to deliver real value.

The critical question becomes: What truly defines value in the context of customer-focused expertise?

Defining Valuable Resource Attributes vs. Job History

Instead of equating value with job titles, organizations should prioritize the following core attributes that consistently drive positive outcomes:

- Deep industry and subject matter knowledge.
- Demonstrated expertise with measurable results.
- Practical know-how related to specific tasks or activities.
- A growth mindset embracing learning and adaptability.

- Strong problem-solving capabilities.
- Genuine empathy and curiosity.
- Collaborative spirit and active listening skills.
- Conscientiousness paired with responsiveness.
- Service orientation and adaptability.
- Effective communication abilities.

These qualities form the foundation of trusted, high-impact resources, not just their résumés.

Aligning Hiring and Resource Allocation with Customer-Centric Mindsets

As outlined in the foundational principles of customer-centric problem-solving, success depends on embracing seven essential mindset shifts:

1. Visualize both current realities and desired future outcomes.
2. Cultivate deep curiosity about customer needs and context.
3. Adopt multiple perspectives to broaden understanding.
4. Lead interactions with empathy to build trust.
5. Pursue thoughtful, data-driven analysis.
6. Remain decisively action oriented.
7. Serve as a trusted guide through complexity.

The Imperative Question: What Are We Really Looking For?

To align resources effectively with evolving customer demands, organizations must move beyond surface credentials. The priority is identifying individuals who embody these valuable attributes and mindsets alongside their knowledge, those who can truly deliver solutions and foster meaningful customer relationships.

This shift requires vigilance and intentionality in recruitment, development, and deployment strategies. By focusing on what truly matters, capabilities and character, organizations will unlock greater customer value and drive sustained success.

Ultimately, redefining value away from mere job history toward authentic expertise and adaptive qualities is critical for any organization committed to thriving in a customer-centric marketplace. The challenge now is to implement this understanding decisively and consistently.

THE COMMERCIAL INSIGHT STRATEGIST: CREATING A NEW ROLE THAT MAKES STRATEGIC ACCOUNT SELLING WORK BASED ON SKILLS AND ATTRIBUTES

Enterprise sales teams face increasing pressure to identify new growth opportunities, navigate complex stakeholder environments, and deliver strategic value while still meeting their targets. However, in many organizations, account executives are expected to manage all these demands alone, often without the necessary tools, time, or support structures to succeed. This challenge inspired the creation of a new role: the Commercial Insight Strategist (CIS).

Role Overview

Investigator

The CIS uses advanced research techniques to analyze data and uncover significant patterns.

Interpreter

The CIS indentifies and maps out untapped growth opportunities for clients.

1 ──────── **2** ──────── **3**

Reporter

The CIS converts data into persuasive narratives that drive strategic decisions.

9.1: Role Overview

Designed from the ground up to translate research into relevance, the CIS enables account teams to move faster, engage more effectively, and pursue meaningful deals. While the account manager focuses on cultivating customer relationships, the CIS enhances these relationships by providing insight, clarity, and early momentum in the sales pursuit.

Key Responsibilities

1 **Insight discovery**
Analyze data and uncover significant patterns to inform strategic decisions.

2 **Narrative building**
Convert data into compelling narratives that drive strategic initiatives.

3 **Opportunity mapping**
Identify and map untapped growth opportunities for key accounts.

4 **Collaborative strategies**
Collaborate with account managers to turn insights into actionable strategies.

5 **Continuous learning**
Stay updated with industry trends and continuously learn to stay ahead.

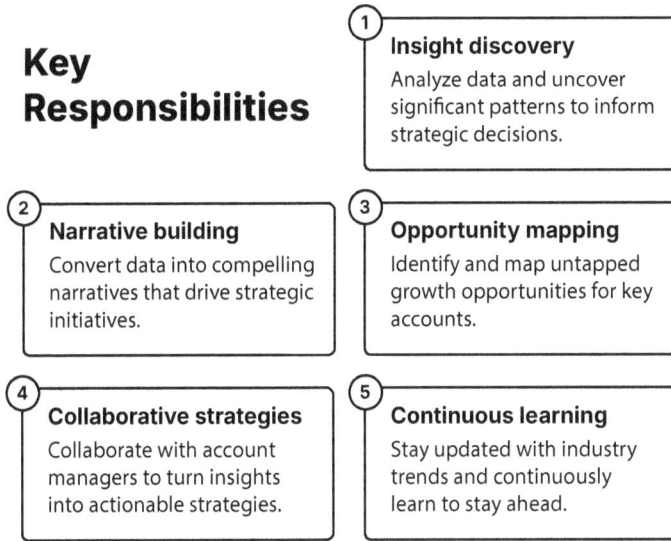

9.2: Key Responsibilities

To establish this role, extensive efforts were made to operationalize the CIS across real accounts, yielding surprising insights. Initially, there was no established playbook or blueprint—only a shared understanding that the traditional account management model was inadequate and needed to evolve. Early obstacles encountered were often internal rather than customer related. Account teams were accustomed to moving quickly, pitching early, and rapidly focusing on solutions—not due to poor intent, but because that approach was ingrained as the winning method.

Leading with solutions prematurely was found to be detrimental, causing teams to miss deeper insights, larger deals, and vital customer-centric focus. The CIS approach requires a different mindset, prioritizing *curiosity* and *critical thinking* as fundamental personal skills and attributes. Instead of following checklists, CIS professionals follow nuanced signals, "breadcrumbs" that indicate change, and persistently investigate until uncovering the patterns that truly matter.

Required Skills and Attributes

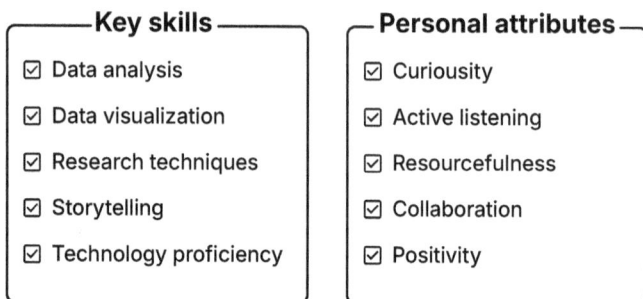

─── Key skills ───	─── Personal attributes ───
☑ Data analysis	☑ Curiousity
☑ Data visualization	☑ Active listening
☑ Research techniques	☑ Resourcefulness
☑ Storytelling	☑ Collaboration
☑ Technology proficiency	☑ Positivity

9.3: Required Skills and Attributes

One illustrative case involved a global enterprise account with a complex organizational structure. The account team was well prepared and engaged in an expansion pursuit with an existing relationship, multiple presentations, and a clear understanding of opportunities. Yet something felt misaligned. A CIS joining the pursuit late noticed a minor detail in a press release about a leadership change within one division. Investigating further, the CIS discovered a budget reallocation linked to an upcoming audit—a fact unnoticed by the team. Consequently, the existing pitch, focused on innovation and growth, was fundamentally misaligned and headed toward failure.

The CIS then reframed the approach, shifting the narrative from "showing new capabilities" to "helping the customer prepare for upcoming challenges." The CIS restructured the engagement plan, identified stakeholders concerned primarily with risk and compliance rather than technology, and helped the team present a message that genuinely aligned with the customer's internal priorities. This reframing, with the same team and offering but viewed through a different lens, led to a larger deal and accelerated progress.

This pursuit marked a turning point, not due to its final outcome but because of early course correction and the role of the CIS in detecting it. The pattern repeated: Traditional account teams missed critical signals because of their focus on speed and product selling, whereas the CIS was trained to pause, zoom out, and reframe the engagement, surfacing risks or opportunities early enough to influence the pursuit's trajectory.

The CIS represents a unique thinker: not reactive, but investigative; someone who identifies shifts before commitments lock the pursuit on a problematic path.

As teams consider introducing the CIS role, key questions arise:

- Will the CIS be dedicated to the role or stretched too thin to detect subtle signals?
- Is there space for the CIS to pursue faint leads, or are they overwhelmed with tasks?
- Can the CIS pause and reframe a pursuit even if this slows momentum?
- Are teams open to hearing and acting on uncomfortable insights?
- Is value measured by influence, insight, and trajectory, not solely closed deals?

Starting with a single, well-placed CIS possessing the right mindset and operational freedom can transform outcomes. Success depends on recognizing what to focus on and what to release.

The need was widely acknowledged. That's why the effort didn't stop at defining the role. A method was created to identify it.

Consider the fact that account sellers are too overwhelmed already:

> *72 percent of sellers feel overwhelmed by the skills they are told they need.*
>
> *85 percent of sellers want leaders to clarify the required skills for success.*[29]

Cracking the Code: How a Smart Self-Assessment Quiz Is Revolutionizing Hiring for a Brand-New Role

Once the impact of the right CIS in the field became clear, the next question arose: How to find more of them? That's where the assessment plays a crucial role.

DO YOU THINK LIKE A COMMERCIAL INSIGHT STRATEGIST?

This short snapshot offers a glimpse into how we assess CIS candidates. It goes beyond just experience to reveal how someone thinks, listens, solves, and adapts in the moment.

---

29 "Supercharge Your Sales Productivity: How to Improve Seller Skills: Optimize Productivity and Increase Results across Your Entire Team," *Gartner*, accessed September 21, 2025, https://www.gartner.com/en/sales/trends/new-high-performing-seller?utm_term=1725639525&utm_campaign=SM_GB_YOY_GTR_SOC_SF1_SM-RM-GBS-SLS&utm_source=linkedin&utm_medium=social&utm_content=Gartner.

Instructions: For each scenario, choose the response that best matches how you naturally operate.

1. You're reviewing three conflicting reports about a customer's strategic direction. What's your next move?
   a. Dig in: Understand the root of the conflict, then share a quick synthesis with the team.
   b. Flag the contradiction and recommend we hold off until we can get clearer direction.
   c. Choose the most recent report and move forward; it's probably the most accurate.

2. A stakeholder just shifted their position on a key initiative. How do you respond?
   a. Reevaluate what may have changed in their world and adjust the approach accordingly.
   b. Let the account manager know; it's their relationship, and I don't want to overstep.
   c. Ask directly what caused the shift so I can update our tracking or notes.

3. You're asked to present insight to a cross-functional team. You . . .
   a. Build a short narrative that ties the insight to business risk, outcomes, and urgency.
   b. Share the data and offer a few options for interpretation, and let the team decide.
   c. Send a summary in advance and focus the meeting on logistics or next steps.

4. A team member forwards an article about a new regulation that might affect a key customer. You . . .
   a. Analyze the ripple effect it could have across the customer's priorities and budget.

    b. Note it, but wait to see if the customer brings it up first.

    c. Archive it—our legal or compliance team will surface anything urgent.

  5. You're juggling multiple priorities, and a new request comes in. It's not urgent, but it's interesting.

    a. Triage based on value: If it helps the pursuit or unlocks insight, I'll carve out time.

    b. Ask for more clarity before investing time; it could be a distraction.

    c. Defer until my plate clears; there's just too much going on.

Scoring Insight

- **Mostly A's:** You think like a strategist: curious, pattern driven, and outcome focused. You're likely a strong fit for the CIS role.
- **Mix of A's and B's:** You show promise. With the right training and structure, you could grow into the role.
- **Mostly C's:** You may be more aligned to supporting roles that benefit from CIS outputs, but don't require the same style of insight synthesis or navigation.

OPTIMIZING COMMERCIAL INSIGHT STRATEGIST RECRUITMENT: A DATA-DRIVEN COMPETENCY ASSESSMENT FRAMEWORK

To ensure the selection of highly performing CISs, a comprehensive Business Manager Assessment Tool was developed, targeting ten critical success attributes essential for the role's demands. These attributes were meticulously defined and structured to assess both technical aptitude and interpersonal effectiveness, forming a robust evaluation framework:

1. **Analytical Skills:** Demonstrates ability to sift through complex, diverse datasets to rapidly identify and qualify actionable insights essential for delivering timely, high-impact intelligence.
2. **Technological Proficiency:** Demonstrates quick adoption and effective use of evolving research tools and CRM platforms, underscoring adaptability to digital environments.
3. **Communication Skills:** Exhibits clarity and persuasiveness when presenting data-driven findings, tailoring messaging to varied audiences, and leveraging multiple communication channels.
4. **Problem-Solving Ability:** Embraces innovative, curiosity-driven solutions anchored in a growth mindset, tackling challenges with tenacity and creative thinking.
5. **Collaborative Attitude:** Actively contributes in team settings, identifying early adopters and leveraging success stories to drive wider organizational engagement and buy-in.
6. **Adaptability:** Responds agilely to organizational shifts in strategy, tools, or team dynamics, maintaining performance amid change.
7. **Client-Centric Focus:** Prioritizes understanding and internalizing client needs, developing personalized insights that address specific challenges, and enhancing client relationships.
8. **Sales Acumen:** Aligns analytic capabilities closely with sales processes, supporting strategic sales initiatives through relevant and actionable insights.

9. **Continuous Learning:** Demonstrates a proactive commitment to upskilling and professional growth, ensuring ongoing relevance and effectiveness.
10. **Leadership Potential:** Exhibits initiative and the ability to influence and guide teams toward shared objectives, signaling future leadership capacity.

This assessment framework is applied during recruitment, complemented by self-evaluation and leadership reviews, enabling a nuanced understanding of each candidate's competencies. Insights gained have revealed that exceptional CISs often come from diverse professional backgrounds and varying levels of direct experience. Notably, success correlates more strongly with demonstrated skills, attributes, and alignment with client account executives' needs rather than traditional industry tenure.

This data-driven, flexible approach underscores the importance of holistic candidate evaluation over rigid background criteria, fueling a dynamic talent pipeline that consistently delivers strategic insights and enriches client partnerships. Organizations are thereby advised to recalibrate recruitment models to prioritize critical competencies and adaptive potential, enabling sustainable commercial impact in an evolving marketplace.

UNLOCKING GROWTH: HOW ALIGNING INTELLIGENCE WITH ACCOUNTS DRIVES REAL RESULTS

Having established a data-driven framework for recruiting CIS who demonstrate the critical competencies to generate impactful insights, it is equally vital to integrate these capabilities effectively within account teams. The true commercial value of a CIS lies not only in their individual skills but in how their insights align with and support broader account strategies. By strategically positioning CIS and other key roles around specific customer needs and solution types, organizations can optimize resource allocation, reinforce client relationships, and significantly increase win rates. The following section explores best practices for aligning intelligence and talent with customer accounts, ensuring that the insights uncovered by high-performing CIS professionals translate into actionable growth opportunities.

Key Steps for Aligning Resources to Accounts:

- **Develop Role Overviews for Account Team Members:** Clearly articulate high-level client-facing roles aligned with the relationship rhythm and engagement complexity. As buyer engagement frequency and seniority increase, the demand for specialized, customer-facing roles grows to support strategic cocreated solutions. Construct detailed role charts that map these roles to the customer's executive buying network to ensure appropriate coverage.

- **Define Specific Responsibilities per Role:** Outline role responsibilities starting with customer expectations and internal workflows. Incorporate support tools and processes and model responsibilities on the best performers to ensure clarity and consistency. This fosters accountability both internally and externally.

- **Specify Required Skills and Attributes:** Enumerate the essential skills and attributes for each role. Develop criteria to identify potential team members from résumés, LinkedIn profiles, and communications. Recognize that high performers often emerge from nontraditional paths, and latent potential should be actively sourced.

- **Create a Comprehensive Inventory of Account-Facing Personnel:** Catalog existing employees by role, including their skill levels, relationship engagement history, and performance metrics. This inventory establishes a foundational resource pool to match personnel with appropriate deals, considering buyer profiles and solution requirements.
- **Group Roles by Customer Needs and Solution Types:** Align team roles with specific customer solution categories to ensure resource availability during proposal stages. Demonstrating aligned, well-rounded teams to customers is crucial—they seek assurance on who will serve them, the responsibilities assigned, and how the team integrates with their organization. This alignment directly impacts win probability.
- **Track Employee Engagements to Optimize Resource Allocation:** Continuously monitor where team members are deployed. Top performers tend to be overbooked; proactive resource management includes mentoring new talent and rotating individuals across engagements to enhance capability and avoid burnout.
- **Implement Systematic Matching of Employees to New Opportunities:** Employ structured, data-driven processes, potentially augmented with AI, to match talent profiles accurately to opportunity demands, including backup and alternative assignments. Proactively aligning resources with customer expectations mitigates the risks of dissatisfaction and account loss.
- **Recruit Proactively Internally and Externally:** Expand

recruitment efforts by targeting candidates who may not have previously considered customer-facing roles but possess the desired skill sets and attributes. Leveraging a skills-to-role matching framework broadens the talent pool, enabling better resource-customer alignment and fostering future growth.

Failure to align resources with these strategic principles risks diminished customer satisfaction and potential account loss. Organizations adopting this disciplined, insightful approach can expect elevated customer engagement, stronger account growth, and improved competitive positioning.

This methodology transforms resource management from reactive assignment to proactive, intelligence-driven orchestration, empowering teams and delighting customers simultaneously.

ROLE BY ROLE: STEPS TO SHAPE THE FUTURE
For Account Teams

Take the initiative by critically assessing the current makeup of your account teams against the ideal structure needed to meet and exceed evolving customer expectations. Identify the gaps between what your customers truly need and the resources currently at your disposal. This exercise is more than a staffing review. It is a strategic analysis of value creation. Imagine the tangible benefits your company could realize by diversifying, augmenting, or optimizing your team's capabilities. Where would you prioritize deploying these enhanced resources? More importantly, can you confidently draw a direct line from these changes to concrete business outcomes such as revenue growth, improved profit margins, or stronger customer loyalty? By embracing this proactive mindset, you position your teams not just as service providers, but as essential drivers of commercial success.

For Marketing

Your company's brand story is incomplete without spotlighting the expertise and passion of the people behind your products and solutions. While products often take center stage, the true differentiator lies in the talent that crafts, implements, and supports them. Step back and envision the full spectrum of your company's capabilities—not just the physical solutions, but the depth of expertise embedded within your teams. How might quantifying and showcasing this human capital transform your customer-facing content? By connecting expert insight directly to the solutions offered, marketing can articulate a richer, more compelling narrative that resonates deeply with customers. This approach unlocks new dimensions of trust and credibility, positioning people and resources as invaluable components of your value proposition.

For Management

People are more than their résumés or job titles. They are dynamic assets capable of unlocking untapped value when viewed through the right lens. Challenge traditional assumptions focused solely on job history and past experience. These criteria often fall short in predicting who will truly connect with customers or deliver differentiated value. Instead, consider the unique skills, personal attributes, and potential of each individual. Those leaders who master this nuanced understanding and craft precise matches between resources and customer needs will decisively outpace competitors. This strategic resource alignment is not just a people function; it is a critical business advantage unlocking a broader and deeper portfolio of customer value.

For Sales Enablement

Sales enablement stands at the crossroads of resource development and solution delivery. Your mission extends beyond training to orchestrating an optimized inventory of resources aligned with your company's solutions. Collaborate closely with human resources to build, nurture, and maintain this talent inventory with an eye toward current and future customer challenges. Ultimately, enablement is charged with maximizing wins, driving account expansion, securing retention, and elevating customer experience, metrics measured by revenue and margin growth. Without your active leadership in defining and sustaining the resource-to-solution matching process, this critical function risks falling through the cracks. Who better to champion this alignment than sales enablement, the function uniquely positioned to bridge talent and business success?

For Commercial Insight Strategists

Your role is both strategic and tactical, ensuring every customer opportunity reflects a laser-focused alignment between identified needs and the resources deployed alongside the recommended solutions. Acting as the steward of the company's capabilities inventory on behalf of account executives, you hold the critical perspective to assess whether the current resource allocation matches the nuances of each opportunity. Where gaps emerge, your insight is vital in recommending supplemental resources or adjusting deployment strategies. Remember, from the customer's point of view, resources are not just support—they are an integral part of the solution itself. Your ability to drive this precision matching directly influences deal success and customer satisfaction, making you a key architect of long-term, sustainable growth.

Now that you have a clear map for aligning the right people—roles, skills, and resources—with your customer's needs, you're ready for the next crucial step: transforming those aligned resources into undeniable value. Resource alignment isn't just about filling seats and checking boxes. It's about enabling meaningful conversations that solve real problems and build trust. In the next chapter, we'll dive into how to communicate that value effectively, moving beyond standard proposals to become true partners in your customer's success journey. Ready to unlock this powerful shift? Let's explore how to turn strategic alignment into lasting impact.

CHAPTER 10:

COMMUNICATING VALUE

There's a profound distinction between merely responding to a procurement-driven RFP or submitting a standard proposal and proactively developing solutions that truly address the heart of a client's challenge. When I shift my mindset from simply selling to genuinely solving, the conversations I have with customer stakeholders change dramatically. It's no longer just about presenting outcomes, capabilities, or resources. It's about collaborating with clients to achieve results that genuinely matter to them.

I have learned, often the hard way, that customers rarely want what I or anyone on my team thinks they want. Early in my career, I built several ROI practices and witnessed firsthand how clients assess value and impact on their own terms. In one case, we developed an ROI calculator and diligently filled it out for the client, only to be told bluntly, "We don't use supplier

ROI calculators." That feedback stung, but it sparked a pivotal insight: If we handed over control and allowed the client to enter their own assumptions, while still providing a robust, fact-based framework, something powerful happened.

We went a step further. A consultant with deep industry expertise guided the client through the methodology—not to persuade, but to educate and empower them to adjust the calculations to reflect their own reality. This turned out to be a breakthrough. Suddenly the client was not only engaged, but they also started requesting this analytical process for every purchase, from every vendor. They wanted a defensible, fact-based foundation for their business decisions, and they valued our transparency and collaborative approach.

Motivated by this success, we embedded an "ROI specialist" into every major account as a trusted adviser. This person mapped business impacts, staged recommendations, and guided clients toward the optimal return. Importantly, this individual wasn't in sales; they served as a catalyst for insight-driven conversations with key stakeholders. The results spoke for themselves: Opportunities grew in both magnitude and depth, fueled by the objectivity, expertise, and partnership we brought to the table.

Reflecting on this experience, I often ask myself: Why aren't we always engaging accounts this way? The blended team approach— combining technical expertise with consultative insight—consistently delivers value that clients recognize and reward. Even more, it has helped me discover countless deeper insights into what customers truly want, and those lessons keep giving, project after project.

Emblaze, the research division of Corporate Visions says on average there's:

> "A 54.5 percent misalignment between how sellers and buyers perceive the core problem to be solved."[30]

If you're in a position to drive change, I encourage you to start by acknowledging that customers often don't know exactly what they need. Don't assume you have all the answers. Instead, provide the tools, frameworks, and expertise that help them gain clarity, supported by your objective, consultative guidance. That's how you build trust and create lasting impact.

PROACTIVE VS. REACTIVE PROPOSALS: ELEVATING YOUR APPROACH TO WIN WITH PURPOSE

Let's be honest: The world of proposals is changing, but many organizations haven't caught up. Traditionally, we respond to RFPs with extensive documents that showcase everything about our company—our achievements, our capabilities, our methodologies. Flip through an average RFP response and you'll find a brief summary of the client's needs, followed by page after page about you. We broadcast these generic proposals far and wide, hoping that volume and polish will generate results.

But here's what we rarely acknowledge: Most of these efforts are wasted. Procurement processes often encourage this rinse-and-repeat approach, but history tells us a hard truth—no one wins an RFP they didn't help shape from the start. By the time an

---

30 Anton Rius, "B2B Buying Behavior in 2025: 40 Stats and Five Hard Truths That Sales Can't Ignore," *Corporate Visions* (blog), January 14, 2025, https://corporatevisions.com/blog/b2b-buying-behavior-statistics-trends/.

RFP is published, the buyer has already made key decisions, and the process is primarily about validating options and checking boxes. In essence, we are reacting, not leading.

Now imagine a different approach. Imagine being the team that moves from the reactive submission of commoditized proposals to deeply proactive engagement. Instead of waiting in line with everyone else, you're working alongside executive buyers at the moment when vision sparks action. When organizations decide to fund an initiative, executives face a complex journey: They must build consensus and secure internal support from multiple stakeholders with varying priorities. In these moments they do not need more boilerplate about your past. What they need are strategic assets that empower their teams to internalize and champion your solution.

This is where you can truly differentiate yourself. Proactive proposals are not about what you offer but about what the client needs to achieve. They frame the path forward with clarity, evidence, and confidence. The executives you aim to serve want to augment their teams with experience and insight—advice grounded in understanding, not self-promotion.

As you craft your next proposal, ask yourself: Are we merely responding, or are we leading the conversation? Are we sharing information, or are we inspiring our buyers to envision and realize a better future with us?

Let's move beyond the comfort of "what we've always done." By becoming true partners—collaborative specialists who empower buyers with actionable road maps, honest assessments, and innovative ideas—we elevate both our proposals and our value. The rewards are undeniable: greater influence, lasting relationships, and a higher rate of success.

Take this opportunity. Challenge your team to rethink not just how you propose, but why. Become the adviser, the problem-solver, and the indispensable ally your clients need. The proactive path isn't just a best practice; it's your strongest competitive advantage.

WHAT CUSTOMERS REALLY WANT: CRAFTING PROPOSALS THAT WIN APPROVAL AND FUNDING

Success hinges on crafting a proposal "pitch" that empowers buyers with exactly what they need to champion, approve, and fund your initiative. Each artifact below serves a strategic function in advancing your cause, reducing barriers, and inspiring confidence among stakeholders.

Branded Identity: Speaking the Customer's Language

Every initiative requires a compelling identity that fits seamlessly within the customer's context. Organizations describe their challenges and opportunities using their own language. Embrace those terms—carefully listen and mirror their internal vocabulary when forming your proposal. When a project name is rooted in the customer's world, it signals deep understanding and partnership. Pipelines that reflect customer-driven naming consistently outperform generic ones, resulting in higher close rates and stronger engagement. The opportunity to help "name" the initiative is more than semantics; it forges authentic alignment and positions your team as trusted advisers.

Conceptual Model: Uniting Stakeholders Around a Shared Vision

A clear conceptual model enables internal advocates to explain how your solution integrates with existing people, processes, information, and technology. Visual diagrams help decision-makers see that your proposal fills gaps, avoids overlap, and is practical, addressing concerns from every angle. These models allow for deep dives into the aspects most critical to each stakeholder, from technology to process improvement. Resistance drops significantly when technical teams see that your solution fits their architecture. By comparing the current state and the future vision, the model illustrates both a solution and a transformation story that stakeholders can see themselves in.

Desired Outcomes: Making Value Tangible

Stakeholders across the organization must see how the future state will benefit them. Rather than relying on exhaustive ROI models, frame clear, understandable outcomes aligned with their priorities. High-level outcome mapping resonates more than spreadsheets, guiding stakeholders' attention toward real, shared wins for their teams and the organization. Chapter 3 unfolds a comprehensive exploration of outcomes.

Outcome Wheel

1 Achieved end state
What does their world look like after working with you?

2 Executive owner
Who is most responsible for making the end state vision happen?

3 High-level initiative
Clients fund initiatives—they don't create budgets for projects. Which type of initiative does the end state align with?

4 Measureable result
What similar examples of success can you share?

5 Impacted stakeholders
Who are the people involved, and how can you help the executive owner manage them?

6 Over a lifecycle
How do you show benefit realization over different periods to help the customer build on small wins to gain momentum?

10.1: Outcome Wheel

ROAD MAP: ILLUMINATING THE PATH FORWARD

A well-crafted road map is more than a timeline—it's a story of progress. Use visual charts to highlight each major phase, offering options to explore greater detail. Transparent, proactive communication of progress fosters trust and sustains momentum. Sharing updates online or in regular check-ins reassures the buying team that milestones are being achieved as promised. When status is clear, anxious interruptions are replaced by anticipation and partnership.

Customer Skills Required: Reducing Barriers

Clarity about the customer's role in success is essential but often overlooked. By specifying the necessary skills and resource commitments, stakeholders can select the right people, plan training if needed, and allocate responsibilities realistically. The less burden placed on the customer, the more attractive the proposal becomes, fueling adoption and accelerating approval. Any objections around "it's too hard" must be surfaced early and addressed directly, ensuring buy-in from those who control resources.

Technical Schematic: Preempting IT Roadblocks

Technical compatibility is the gatekeeper to progress. A detailed schematic shows how your solution integrates with the customer's existing systems, validating that all requirements are met and common IT objections have been anticipated. Preparing these diagrams removes friction, reduces delays, and reassures technical stakeholders. Involving IT representatives from the start allows you to proactively address concerns, building confidence that your offer is ready for seamless adoption.

Continuous Value: Sustaining Trust and Engagement

The value of a solution must withstand changes in personnel and priorities. By establishing regular checkpoints and validating impact at each stage, your team demonstrates commitment beyond the initial sale. Capturing feedback and sharing results across the customer's organization amplifies success and builds momentum for future opportunities. Utilizing shared digital engagement portals can further reinforce this ongoing partnership, making value visible and trust enduring.

By meticulously addressing each of these areas, the proposal becomes more than a pitch—it transforms into a blueprint for shared success. This is how initiatives earn internal advocacy, withstand scrutiny, and achieve the approvals needed to move forward. Seize the opportunity to differentiate by equipping your customers to be heroes within their organization, and lasting partnerships will follow.

PRIORITIZE RISK REDUCTION TO SECURE BUY-IN

Risk avoidance is a powerful motivator for customers, often taking precedence over the pursuit of new opportunities and benefits. Customers are fundamentally oriented toward escaping pain and minimizing any factors that could expose them to blame. Effective advisers recognize that, from the customer's perspective, every new solution presents a spectrum of risks that must be identified, evaluated, and methodically addressed.

Success begins by adopting the customer's lens: Systematically quantify and catalog every potential risk that might concern them. Develop a detailed risk checklist that anticipates possible pitfalls, then proactively chart a strategy to mitigate each one. Establishing a robust risk-reduction plan should be the foremost priority. Once concerns are alleviated, the conversation about tangible benefits becomes markedly more persuasive and relevant.

It is a common misconception that decision-making in purchasing is strictly rational. Were this the case, clear-cut benefits would win every time. In reality, purchasing decisions are intricately emotional. Buyers seek not only to achieve a better future but to eliminate perceived threats along the way. This emotional calculus underscores the value of deeply understanding the customer's underlying motivations—what is known as the "job to be done." Requirement gathering should thus be a process of discovery, probing beyond surface-level needs to uncover the hidden drivers: personal stakes, anxieties, desires, and preferences.

The most effective professionals never settle for the obvious answer. They persistently search for the "need behind the need," the unspoken incentives or concerns that propel customers to embrace change. By skillfully revealing and addressing these deeper motivations, it becomes possible to transform risk into trust, skepticism into advocacy, and hesitation into commitment.

Adopt the mindset of a trusted guide. Lead with risk reduction, and genuine value will follow.

ASSESS THE VALUE EQUATION WITH PRECISION

When proposing a solution, even the most impressive benefits can be overshadowed if perceived risk remains high. Often stakeholders hesitate not due to a lack of value, but due to risk outweighing reward. Think of the following as the value equation:

$$\text{(Total Benefit − Cost) / Risk}$$

Regardless of the benefit presented, elevated risk, measured by any metric relevant to the decision-maker, quickly becomes the primary barrier to progress. A compelling solution must do more than promise results; it must **actively minimize risk**. Success is found in meticulously identifying and eliminating obstacles, consistently reinforcing the reliability of the proposed solution. Where proven outcomes and a track record of delivery exist, those become foundations for credibility. Ultimately, progress hinges not on benefits alone, but on setting buyers at ease by removing uncertainties.

STAY ATTUNED TO TIMING SIGNALS

"We aren't ready yet," a frequent objection, is itself a symptom of risk aversion. Attempting to rush or override such reservations rarely yields positive outcomes and may erode trust. Instead, adopting an advisory stance leverages patience as a strategic asset. Guiding the conversation in a consultative manner positions the adviser as a trusted driver of the initiative. Codifying buyer readiness or timing into sales processes can further sharpen pipeline accuracy. By proposing engaging but flexible steps, the process can be extended as necessary to alleviate fears of organizational disruption. This approach demonstrates empathy toward client pacing while maintaining momentum.

PRACTICE ADAPTIVE ALIGNMENT FOR LASTING RELEVANCE

The landscape surrounding every customer is in a state of continual evolution. Demonstrating the capacity to adjust solutions in response to shifting circumstances is a hallmark of a strategic partnership. Remaining relevant means staying agile and being willing to realign offerings as new factors emerge. Regularly revisiting and, if needed, refining proposals in response to client feedback keeps engagement high and showcases an active investment in their ongoing success. This proactive flexibility is often met with genuine surprise and appreciation from clients, leading to deeper connections and greater trust. CISs excel by continuously monitoring client contexts and suggesting meaningful adaptations, ensuring that proposed solutions not only meet current needs but also anticipate future ones.

THE PATH FORWARD

True impact comes from mastering the interplay of value, risk, and timing while maintaining a nimble posture. By prioritizing risk mitigation, embracing patience, and championing flexibility, sellers and advisers inspire deeper client confidence and position themselves as indispensable partners in organizational progress.

ROLE BY ROLE: STEPS TO SHAPE THE FUTURE
For Account Teams

Effective proposals begin by working backward from the customer's expressed challenges, goals, and objectives. It is crucial to articulate these insights clearly in the customer's own language, demonstrating deep understanding and alignment. Incorporate visual aids that illustrate the cocreated solution from the customer's perspective, showing how it seamlessly integrates into their environment to lower risk. When engaging stakeholders with specific concerns, provide detailed drilldowns that compare the current state to the desired future state, highlighting how risks or issues are mitigated. Maintaining a sharp focus on aligning the solution with the customer's evolving needs and continuously refining the approach will build trust and accelerate success.

For Marketing

Systematic assessment of proposal formats to identify the most effective ones is essential. Develop a centralized library of templates, schematics, and road maps that sales teams can adapt efficiently. This not only saves time but ensures consistency and quality across engagements. Recognize that creativity varies among teams; some will innovate effortlessly while others require structured support. Elevate the professionalism of visuals and consolidate all common pitch content in a single, accessible repository. Proactively monitor and share top-performing materials and establish feedback loops by regularly soliciting input from account sellers. Encouraging them to document customer reactions within the system enables continuous refinement and enhances overall impact.

For Management

Adopting a front-office pitch team with diverse specialized skill sets transforms deal outcomes. Incorporating subject matter experts alongside sellers instills greater confidence in customers, which directly correlates with improved close rates and larger deal sizes. These specialists provide critical insights and expertise that customers deem vital to success. Management must prioritize identifying, tracking, and deploying these valuable resources efficiently. When managed strategically, their involvement elevates the credibility and effectiveness of solutions, driving superior business results.

For Sales Enablement

Introducing external experts specializing in requirements gathering and executive buyer engagement can significantly enhance seller performance. Many account sellers face challenges elevating conversations to executive levels and managing multi-specialist collaboration. Bringing in skilled consultants or trainers to provide tailored guidance bridges this gap, empowering sellers to engage more confidently and strategically. This investment accelerates sellers' growth, refines their approach, and increases the likelihood of securing executive buy-in, resulting in more compelling deals.

For Commercial Insight Strategists

Constantly reevaluate the alignment between the proposed solution and the customer's expressed needs. Go beyond surface-level requirements—dig deeply to uncover the "need behind the need." This analytical rigor is the foundation for delivering precise value and reducing risk. Act as a strategic analyst alongside the account executive, rigorously breaking down the customer's problem and evaluating how effectively the solution mitigates risk. By maintaining this disciplined focus, the CIS ensures the solution fits perfectly, fostering confidence and driving customer satisfaction.

You've seen how shifting from reactive selling to proactive partnership transforms conversations, builds trust, and unlocks real value. But what happens when that trust meets the challenge of sustaining engagement every single day? How do you keep ideas flowing, ownership shared, and momentum alive, especially when your client's expectations feel insatiable or even unreachable?

That's exactly where the next chapter takes us: the art and science of creating dynamic, transparent engagement that turns tough demands into opportunities for deeper connection and collaboration. Ready to explore a game-changing tool that redefines ongoing client relationships and powers continuous success? Let's dive in.

CHAPTER 11:

BUILDING A COMMERCIAL COMMUNITY

When the account team told me the CMO of a major retailer demanded, "I want a new idea on my desk every day," their frustration was clear. They found the CMO difficult. No matter how many ideas they presented daily, he wouldn't even look at them. I understood their exasperation, but I also realized something important: Our team's focus on tactical marketing and advertising campaigns didn't match the strategic mindset from which the CMO was operating. It was doubtful that the ideas generated would truly resonate or meet his expectations.

Reflecting on this challenge, I saw an opportunity beyond simply delivering ideas daily. The real value lay in creating a structured, accessible repository where every idea we shared was stored, organized by topic, and available for review. This would provide undeniable proof of our ongoing engagement and commitment. No longer could the CMO claim we weren't delivering. There would be a clear history and transparency. The question was how to make this concept tangible and user friendly. This moment of insight sparked the creation of what we would later call the "client engagement suite."

Within three months, we designed and launched a new customer-facing portal, which quickly became a cornerstone of our customer account communications. Central to this portal was the Idea Center—a dynamic space dedicated to housing every idea we provided daily to the CMO. Although initially driven by specific customer demand, the Idea Center revealed its profound importance: It extended far beyond simple idea delivery. It became a strategic asset that enhanced our credibility, fostered continuous collaboration, and ensured alignment with the customer's evolving priorities.

From this experience, I learned that truly understanding the customer's operating level and establishing transparent, documented engagement mechanisms can transform a difficult relationship into a powerful partnership. The Idea Center concept exemplifies how attentive listening paired with innovative solutions can not only meet but exceed customer expectations, turning challenges into lasting success.

ELEVATING CUSTOMER RELATIONSHIPS: THE BLUEPRINT FOR MODERN ENGAGEMENT

Building robust customer relationships has never been more critical, nor more complex. Chapter 3 dissected the key drivers of stakeholder perception:

- **Accessibility:** Is partnering with your organization seamless and efficient?
- **Responsiveness:** Are customer needs met promptly and thoughtfully?
- **Performance:** Do solutions consistently deliver on high expectations?
- **Outcomes:** Are shared objectives not just met, but exceeded?

Today's landscape, defined by AI and rapid digital transformation, demands a shift from passive interactions to proactive value creation. Engaging early in the customer's buying cycle, cocreating tailored solutions, sets exceptional organizations apart. Offering customers transparent access to a dynamic library of shared ideas and enabling real-time collaborative ideation fosters true partnership and innovation.

The Amazon example is instructive. When Jeff Bezos founded Amazon in 1994 with a visionary belief in e-commerce's potential, it wasn't about selling products online; it was about removing friction and reimagining convenience. This relentless focus is on a frictionless customer experience, redefining industries. B2B leaders must embrace this mindset, particularly for their most valued customers, ensuring every interaction is intuitive, accessible, and forward looking.

Alignment is essential. Bridging internal strengths and customer engagement through modern digital channels creates a sustainable competitive advantage. The era of reliance on traditional tools—email and phone calls—is fading. Strategic relationships now require robust, digital-first platforms that mirror the immediacy and convenience customers expect from every digital interaction.

Organizations willing to adapt, innovate, and fully align with customer needs will build relationships that thrive. The lesson is clear: Embrace digital modernization, prioritize seamless engagement, and commit to a future of shared success.

DESIGNING THE EXPERIENCE: A STAKEHOLDER-CENTERED APPROACH

Effective interface design begins by working backward from the stakeholders' unique needs. Initially focusing on the CMO, the design prioritized a central Idea Center as a hub for innovation. Recognizing the CMO's strategic role, relevant industry trends and analyst reports were integrated to provide critical market insights. Complementing this, an overview of the team's capabilities tailored to retail was added, along with a request-response form, competitive analysis, and upcoming event information, creating a robust, value-driven experience.

However, a singular focus on the CMO overlooks the diverse needs of other stakeholder roles within the account. Each role demands distinct information and functionality to perform effectively. This insight led to the development of tailored interfaces for multiple stakeholder groups:

- **Executives:** Idea Center, industry trends, analyst reports, competitive analysis, upcoming events, and team overviews.
- **Procurement:** Contracts, renewal statuses, contacts, pricing, and billing information.
- **Campaign Managers:** Campaign schedules, performance metrics, objectives, business impacts, and KPIs.
- **Data and Technology Teams:** Infrastructure status,

uptime, security alerts, contacts, tickets, evaluations, and data inventories.

- **Marketing Program Leads:** Historical offers, audience metrics, segmentation data, competitor programs, and results.
- **Analytics Teams:** Model performance, training schedules, recommendations, and development statuses.

All relevant outputs were already delivered to customers through disparate channels. The challenge and opportunity were to consolidate these assets into a unified, accessible online platform customized for each stakeholder's specific user interface requirements. This approach not only enhances usability but also drives engagement by ensuring each stakeholder receives the precise insights and tools needed to excel.

Designing the experience by working backward from the stakeholders creates a strategic, efficient, and impactful solution, one that transforms data and resources into actionable intelligence tailored for every role in the organization.

The Critical Advantage of Internal Organization and Asset Standardization

Imagine spreading every customer-related asset across an empty room, files, documents, and communications, scattered in disarray. Much like moving house, where accumulated clutter and forgotten items resurface, corporate account assets over the years often become decentralized, fragmented across hard drives, shared folders, archived emails, and chat logs. This lack of consolidation is not just inconvenient; it fundamentally undermines efforts to deliver a seamless, modernized customer experience.

Decentralization creates barriers: Your customers face inconsistent, difficult-to-access materials, leading to repeated information requests via emails and calls. This inefficiency wastes time and erodes customer confidence, reflecting poorly on your organization's professionalism and agility.

The transformation begins by recognizing the value and challenge of centralizing these assets. Initial resistance from internal teams is natural, rooted in discomfort with change and an inability to envision the improved future state. Yet moving past this phase unlocks a powerful outcome: the creation of a cohesive digital commercial ecosystem. This streamlined environment fosters transparent communication, enhanced collaboration, and faster resolution times, building a foundation that delivers lasting strategic benefits and measurable ROI.

Prioritizing internal organization and standardization of account assets is not merely a tactical step; it's a strategic imperative for organizations aiming to thrive in a competitive landscape. The return is a modernized, scalable customer experience that drives loyalty, efficiency, and growth. Embracing this change today sets the stage for sustainable success tomorrow.

THE VISION OF A STRATEGIC COMMERCIAL COMMUNITY FOR ACCOUNT TEAMS AND CUSTOMERS

A strategic commercial community is a customer-centered ecosystem grounded in purposeful value exchange, collaboration, and innovation. It envisions account teams and customers as equal partners who jointly identify and deliver meaningful business outcomes aligned with strategic priorities.

This community transcends traditional roles: Account teams

act as strategic partners, deeply embedded in the customer's success journey, while customers engage actively as contributors, shaping discussions and cocreating solutions. This dynamic fosters co-ownership and mutual responsibility among all stakeholders.

The true value lies in strengthening strategic relationships through open dialogue, shared experiences, and transparency. Such a community drives greater loyalty, advocacy, and resilient business outcomes. By nurturing this collaborative network, account teams deepen their relevance, unlock growth, and achieve sustainable competitive advantage in the evolving B2B landscape.

> *When B2B firms create mature customer advocacy/ community programs, they can expect +23 percent retention, +21 percent revenue from existing customers, and +22 percent competitive wins.*[31]

BUILDING A DYNAMIC COMMERCIAL COMMUNITY

Building a thriving commercial community requires purposeful design and ongoing engagement. Early and continuous involvement of customers across the buying journey fosters cocreation of solutions and reinforces strong partnerships.

Modern commercial communities leverage digital platforms to enable flexible, real-time interaction beyond traditional in-person events. The core of this environment depends on providing stakeholders with timely, relevant, and actionable insights, ensuring consistent value that motivates active participation.

Effective engagement incorporates concise, visually compelling

---

31 "Prioritize Postsale Customer Marketing to Drive Business Value and Growth," Forrester, accessed September 21, 2025, https://images.g2crowd.com/uploads/attachment/file/176681/ Influitive-Forrester-Prioritize-Postsale-Customer-Marketing-Report.pdf.

content such as brief videos to keep members informed and connected. This proactive, always-on mindset shifts account management into a strategic advantage, embedding collaboration, agility, and transparency into everyday interactions.

While building this dynamic ecosystem entails challenges, it offers forward-thinking organizations immense opportunities to deepen relationships and accelerate growth.

THE TEN ESSENTIAL STEPS TO BUILDING A THRIVING DIGITAL COMMERCIAL COMMUNITY

Success in today's digital landscape requires deliberate, structured, and adaptive strategies for engaging and growing key accounts. Building a thriving digital commercial community transforms passive customers into active partners and amplifies growth across every touchpoint. Mastery of these ten steps provides a practical blueprint to elevate account experiences, foster collaboration, and drive competitive advantage.

1. Centralize Account Assets

Consolidate all account-related materials, including documents, communications, and resources, into a unified, easily navigable repository. Use systematic directories and consistent categorization to streamline access and minimize confusion.

- Standardize formats and branding to ensure materials reflect the highest quality and project a strong, unified brand image.
- Establish clear update schedules so the latest assets are always available.
- Leverage content management systems to facilitate efficient

sharing via links, significantly reducing time wasted in manual information retrieval.

- Implement automated alert systems to notify stakeholders of updates instantly, establishing expectations for responsiveness and transparency.

2. Design User Interfaces by Stakeholder Role

Develop user interfaces tailored to the unique needs of each stakeholder group.

- Map out interface elements—schedules, updates, contacts, and news—so users can access what matters most to them.
- Enable flexibility, allowing stakeholders to customize layouts for both desktop and mobile.
- Conduct internal testing, gathering team input to refine the user experience before launch.

3. Review Content Through the Customer Lens

Solicit feedback from individuals outside the core account team to assess relevance, clarity, and usability.

- Utilize structured checklists to evaluate the portal's value from roles across your customer's organization.
- Integrate feedback iteratively, continuously closing content gaps and improving usability.
- Position the platform as a branded experience, tailored to each account's culture and objectives, not just as a transactional tool.

4. Align and Mobilize Internal Resources, Then Launch

Engage the account team in a deep exploration of the new interface's potential.

- Gather and integrate their insights to foster ownership and innovation in customer engagement.
- Secure leadership buy-in to enhance cross-functional communication and unify expansion efforts.
- Announce the launch to customer stakeholders through compelling messaging, highlighting organizational commitment to collaboration and continuous improvement.

5. Evaluate Engagement Data for Insight

- Track platform and content usage metrics by stakeholder role to identify patterns and preferences.
- Monitor which materials are most accessed, uncovering what drives real stakeholder interest.
- Analyze comments and feedback to gain actionable intelligence on customer objectives, priorities, and satisfaction.
- Assess the effectiveness of alerts to optimize communication strategies.
- Incorporate quick health check prompts in daily interactions to monitor relationship strength and anticipate challenges.
- Proactively identify both emerging opportunities and underlying risks, enabling timely interventions and tailored solutions.

6. Integrate Insights into Account Strategy

- Directly embed engagement data and customer feedback into account plans, ensuring real-time situational awareness.
- Use these insights to drive adaptive, responsive strategies, closing the loop between action and outcome.
- Recognize that buying networks are increasingly complex. Fast, accurate feedback accelerates decision-making and sharpens the competitive edge.

7. Focus on Continuous Performance Improvement

- Develop actionable performance reports and communicate key metrics across teams.
- Identify improvement opportunities from value propositions and resource allocation to solution delivery.
- Standardize the communication of KPIs to drive more meaningful, data-informed conversations across functions.

8. Incorporate Partners for Holistic Solutions

- Identify gaps your organization cannot fill alone and forge partnerships that complement core capabilities.
- Vet and introduce trusted partners to customers, ensuring seamless, end-to-end solutions.
- Own the process to relieve customers of unnecessary burden, earning repeated trust and forming durable relationships.

9. Become the Dominant Customer Collaboration Hub

- Be proactive in modernizing stakeholder experiences through digital platforms that enable timely, relevant, and actionable engagement.
- Harness organizational efficiencies by integrating standardized content, automated feedback loops, and collaborative digital workspaces.
- The result: remarkable savings and accelerated customer outcomes, leading to measurable improvements in commercial performance.

10. Foster Community One Account at a Time

- Treat every key account as an ecosystem with vast growth potential.
- Embrace real-time updates and seamless team-member collaboration as normal practice, accelerating innovation and responsiveness.
- Extend best-practice sharing across accounts, harnessing network effects for maximum impact.
- Connect executives around common industry challenges to create even broader value, cementing a reputation as an indispensable strategic partner.

In a rapidly evolving digital world, taking structured steps to build a commercial community is no longer optional. It is an imperative. By approaching each account as a fertile ground for continuous growth, embracing technology, and weaving together expertise and partnership, organizations secure their place as leaders, innovators, and trusted advisers in every market they serve.

ROLE BY ROLE: STEPS TO SHAPE THE FUTURE
For Account Teams

Imagine transforming your accounts into vibrant commercial communities where collaboration, innovation, and value creation thrive. Building and nurturing such an environment isn't just beneficial; it's essential. As an account leader, you must champion the integration of marketing, enablement, and management resources to develop digital experiences tailored for your customers. These platforms enable you to streamline daily account management, reduce friction, and uncover new avenues for growth. By advocating for and leveraging these tools, you position your team not only to deliver exceptional service but to accelerate expansion strategies that unlock long-term revenue potential. The future of account management is about fostering relationships at scale—making your accounts more efficient, connected, and strategically positioned for success.

For Marketing

Marketers, picture taking account-based marketing beyond its current boundaries, turning it into a dynamic engine for customer expansion through cross-selling and upselling. Your real power lies in curating relevant, timely content that fuels digital customer engagement portals. This is where marketing's role transforms from simply delivering messages to orchestrating experiences. The content you provide isn't just information. It's the fuel for customer trust and business growth. Furthermore, these engagement hubs generate invaluable insights into customer needs, behaviors, and preferences, enriching not only expansion programs but broader acquisition strategies. Treat your flagship accounts as industry beacons; what you learn from them provides a window into market trends and customer priorities at scale. Leveraging these insights strategically elevates marketing's impact and drives measurable business outcomes.

For Management

Leadership, your mandate is clear but challenging: Systematically grow account revenue and profit margins. Achieving this requires more than skilled people. It demands intelligent systems of insight and collaboration that empower your teams and customers alike. Relying solely on human effort without the support of technology and data-driven frameworks puts you at risk of missed opportunities and weakened market positioning. The competition is racing forward, equipping their account teams with these modern capabilities. Ask yourself, do you want to lead or lag? Investing decisively in digital platforms, tools, and the teams that manage your most valuable accounts is not optional; it is a strategic imperative. The pace of change is relentless. Commit today to building an infrastructure that sustains growth, drives loyalty, and defends your competitive edge.

For Sales Enablement

The traditional mindset has been to simply make it easier for salespeople to sell. Now shift the focus: your mission is to make it easier for customers to buy. This subtle yet profound difference requires creating frictionless buying environments—digital, intuitive, and resource rich. Partner with marketing and account teams to centralize content and assets into seamless customer-facing portals. Consider the daily frustration account teams experience when manually delivering these resources; how much better it would be if customers could self-serve at their convenience. By leading this transformation, you not only reduce friction in the buying process but also empower sales teams to do what they excel at: building relationships and closing deals. Your leadership in this space moves the entire commercial organization closer to customer-centric excellence.

For Commercial Insight Strategists

You hold the keys to understanding customer priorities and challenges like no one else. Your role is to decode what customers want to see, hear, and engage with daily, distinguishing immediate pain points from long-term goals, and tailoring insights to specific roles and decision-makers. Your insights are the currency of customer relevance and business agility. Establish strong collaborative channels with enablement and marketing teams to ensure these insights are delivered precisely when and where they are needed. Alerts, targeted content, and timely communications powered by your expertise create a cadence of engagement that keeps your customers informed, confident, and connected. By mastering this orchestration, you elevate the entire commercial ecosystem, turning insight into influence and opportunity.

As you reach the end of this journey, consider the value waiting to be unlocked—the very "rare gem" introduced in the prologue. All along, this guidebook has invited you to see what was hidden in plain sight: the transformational power of deep, actionable customer intelligence.

Now, with fresh perspective and new tools at your fingertips, the question isn't whether you can change how your teams operate; the real question is how much impact you're willing to create. Every insight gleaned, each digital collaboration, and every seamless customer experience becomes the fuel for a vibrant commercial ecosystem. The baton is in your hands.

So, as you prepare to put these ideas into practice, remember true commercial growth isn't about chasing the next opportunity; it's about building a culture where opportunities find you. Let's move beyond transactional selling and toward cultivating communities of customers who feel known, valued, and energized by your partnership.

The promise described at the outset, a systematic, customer-centric approach that drives lasting success, starts with a single step: embracing curiosity and committing fully to what's possible. The next chapter isn't written yet. It's yours to create. Let's unlock it together.

AFTERWORD:
IT'S TIME TO MOVE

Companies who take action and lead the way in systematically aligning capabilities, resources and value to the needs of the most valuable B2B customers that spend the most will create massive competitive advantage by being first. AI is accelerating the ability to make this systematic alignment a reality very quickly today, and it's outpacing the speed with which organizations can adapt. However, automated workflows, continuously delivering valuable actionable customer insights where and when they are needed to inform constant collaboration with customers to address their shifting challenges is the future.

Companies need start embracing and re-aligning to this future state now. Working backwards from customers to inform how to best optimize AI enabled workflows to meet customer needs and expectations is the best starting point. Organizing and prioritizing action plans aligned to an inventory of customer needs and then unlocking the value of your organization to meet those needs is what customers, employees and shareholders want. Now it is possible to dramatically accelerate this realignment with customers for driving towards a new frontier of profitable revenue growth tied to enterprise accounts.

Contact us, let us know how we can help you activate your path forward, where to start, what to think about and how to drive meaningful impact now while re-aligning workflows, systems, and people over time to achieve competitive advantage before it's lost. The time to *Go to Customer* is now.

APPENDIX 1

VUCA DRIVEN BUSINESS CONDITIONS AND CORRESPONDING INSIGHTS NEEDED BY CATEGORY

B2B ENTERPRISE CONDITIONS	VOLATILITY	UNCERTAINTY	COMPLEXITY	AMBIGUITY
Industry Conditions	**Challenge:** Every market shift is an opportunity for those ready to act. **Action:** Embrace change by empowering your teams with cutting-edge analytics. Set the pace, don't just keep up.	**Challenge:** Uncertainty breeds disruption but also sparks innovation. **Action:** Lead with vision—invest in forward-thinking research to anticipate trends and innovate boldly.	**Challenge:** Complexity can slow competitors, but not agile leaders. **Action:** Turn complexity into an edge with adaptive compliance and data-driven segmentation. Mastery of the landscape sets you apart.	**Challenge:** The unknown rewards the prepared. **Action:** Champion collaboration and future-focused teams; those who read the signals earliest are best positioned to define the industry's direction.

B2B ENTERPRISE CONDITIONS	VOLATILITY	UNCERTAINTY	COMPLEXITY	AMBIGUITY
Competitor Conditions	**Challenge:** Managing market disruptions from abrupt competitor moves, such as strategic pivots, new market entrants/exits, and price wars. **Action:** Continuously monitor and analyze competitors with advanced intelligence tools. Develop adaptive pricing strategies and remain agile to respond swiftly.	**Challenge:** Predicting competitors' future strategies and innovations in an unpredictable environment. **Action:** Institutionalize ongoing competitive analysis, scenario planning, and foster external partnerships to increase visibility and adaptability.	**Challenge:** Navigating the layered strategies, diverse partnerships, and intricate market positioning of rivals. **Action:** Deploy advanced analytics and AI-driven tools to decode complex competitor ecosystems and inform decisive actions.	**Challenge:** Deciphering competitor intent when signals are unclear, data is incomplete, or misinformation abounds. **Action:** Leverage triangulated data sources and multifaceted intelligence gathering to construct reliable insights.
Artificial Intelligence (AI)	**Challenge:** The rapid pace and unpredictability of AI advancements put pressure on traditional operational models. Leaders face constant change. What is cutting-edge today may be obsolete tomorrow. **Action:** Make continuous learning a nonnegotiable team priority. Invest in up-to-date AI training and proactively review emerging AI governance frameworks to build a future-proof workforce. By embracing change, you create an organization ready to capitalize on every wave of innovation.	**Challenge:** The long-term impact of AI on your business models and talent pipelines remains unclear. Decisions made today will reverberate unpredictably—are you prepared for the unknown? **Action:** Conduct scenario planning and regular impact assessments. Adopt flexible, adaptive business models that allow ethical experimentation and swift recalibration. Launch workforce reskilling programs, empowering your people to stay relevant and resilient as AI redefines work.	**Challenge:** Integrating AI into complex legacy systems while upholding ethical standards can overwhelm even seasoned teams. Overlooking interdependencies or compliance gaps risks costly setbacks. **Action:** Map your operational ecosystem thoroughly. Prioritize modular, scalable AI deployments to absorb future changes with minimal disruption. Establish robust ethical oversight, championing transparency and responsibility at every stage.	**Challenge:** The boundaries of AI's potential are blurred. Unclear use cases and evolving guidelines make it daunting to separate real opportunities from hype. **Action:** Advance AI literacy at every level. Encourage cross-functional innovation labs to explore and validate new use cases. Foster a culture of experimentation—where curiosity becomes your competitive edge and ambiguity is a trigger for discovery, not paralysis.

B2B ENTERPRISE CONDITIONS	VOLATILITY	UNCERTAINTY	COMPLEXITY	AMBIGUITY
Digital Transformation	**Challenge:** Rapid shifts in technology threaten to outpace current capabilities and outdated annual planning cycles risk leaving your business behind. **Action:** Establish a culture of continuous improvement. Embrace agile project governance, empower frontline decision-making, and prioritize real-time trend monitoring through strategic partnerships. Make scenario planning a routine discipline to anticipate disruptive events and turn turbulence into innovation opportunities.	**Challenge:** Unpredictable outcomes exacerbate risk, making it challenging to identify and deploy the right digital strategies in the face of incomplete information. **Action:** Conduct digital maturity assessments and launch pilot programs to test and de-risk innovations. Foster a culture of deliberate learning, equipping teams to rapidly pivot based on fresh insight. Prioritize adaptability over rigid processes.	**Challenge:** Coordinating digital transformation across diverse teams and platforms risks duplication, inefficiency, and operational delays. **Action:** Develop a centralized digital strategy underpinned by clear governance, disciplined prioritization, and robust KPIs. Streamline tech investments and operations to focus resources on high-impact value creation; eliminate non-strategic processes.	**Challenge:** Measuring the impact of digital initiatives and forecasting what's next are increasingly difficult amid incomplete data and unclear cause and effect. **Action:** Implement advanced analytics and performance frameworks to extract actionable insight. Promote a bias for action over paralysis by analysis. Make it safe to experiment, fail fast, and continuously refine digital priorities for the business.

B2B ENTERPRISE CONDITIONS	VOLATILITY	UNCERTAINTY	COMPLEXITY	AMBIGUITY
Cyber Security	**Challenge:** The pace and unpredictability of *emerging cyber threats*, including AI-driven attacks and advanced ransomware, can catch organizations off guard, disrupting business operations instantly. **Action:** Empower your teams by investing in advanced threat detection and AI-powered response solutions that learn and adapt as fast as the threats themselves. Make regular system updates and security patches a foundational discipline.	**Challenge:** Anticipating the potential impact of cyber incidents is increasingly difficult in a landscape marked by new threat actors and evolving attack strategies. The real question: Are your defenses sufficient for tomorrow's threats? **Action:** Build resilience. Conduct rigorous, regular cybersecurity audits and scenario-driven risk assessments. Equip your organization with robust incident response and business continuity plans—be ready to contain, recover, and adapt at a moment's notice.	**Challenge:** The proliferation of digital touchpoints, diverse security protocols, and overlapping compliance mandates creates daunting management complexity. One misstep can open a door to cybercriminals. **Action:** Streamline and unify. Embrace Zero Trust architecture, standardize core security protocols, and leverage automation for compliance and monitoring. Foster a culture of cross-functional collaboration to transform complexity into security strength.	**Challenge:** Judging the security of *new technologies* and digital innovations—like Internet of Things (IoT), cloud, or AI—remains a moving target. Unknown vulnerabilities can lurk, posing risks that are difficult to quantify or even see. **Action:** Champion proactive exploration. Adopt industry-recognized cybersecurity frameworks and engage trusted third-party specialists for security reviews. Instill a mindset of continuous learning and skepticism—question, test, and verify every new technology before deployment.

B2B ENTERPRISE CONDITIONS	VOLATILITY	UNCERTAINTY	COMPLEXITY	AMBIGUITY
Cloud Computing	**Challenge:** Rapid-fire changes in cloud services and provider offerings demand that IT leaders stay agile to avoid lock-in or disruption. **Action:** Establish a flexible cloud strategy leveraging both multi-cloud and hybrid cloud architectures. Build adaptability into your technology road map—a dynamic foundation is your shield against turmoil.	**Challenge:** Evolving regulations, security threats, and privacy mandates expose organizations to unforeseen risks. **Action:** Deploy robust encryption, implement AI-driven compliance monitoring, and ensure rigorous, ongoing regulatory training for your teams.	**Challenge:** Migrating legacy systems, managing hybrid environments, and ensuring interoperability often overwhelm even advanced IT teams. **Action:** Harness automated migration tools and comprehensive cloud management platforms to simplify complex transitions. Map dependencies and phase migrations for minimal business disruption.	**Challenge:** The array of cloud service models (infrastructure as a service [IaaS], PaaS, software as a service [SaaS]) creates ambiguity regarding which will deliver the greatest long-term value. **Action:** Build a clear, outcome-focused cloud strategy rooted in measurable business objectives. Prioritize platforms that align with your organization's future vision.
Technology	**Challenge:** Rapid technological advancements and market disruptions are redefining the competitive landscape at an unprecedented pace. **Action:** Proactively cultivate an innovation-driven culture and invest in continuous learning—make technology road maps a living part of your growth strategy. The organizations that adapt swiftly set the pace; those that fall behind risk irrelevance.	**Challenge:** The unpredictable trajectory of emerging technologies creates uncertainty about their impact on core business operations and customer expectations. **Action:** Engage in rigorous technology foresight and robust trend analysis. Develop adaptive strategies that anticipate, rather than react to, external changes—future-proofing your competitive position.	**Challenge:** Integrating breakthrough technologies with legacy systems requires overcoming significant operational and cultural barriers. **Action:** Adopt middleware solutions and modular architectures that allow agile integration—break down silos and enable scalable innovation without disrupting current workflows.	**Challenge:** Navigating the ambiguity of new technology opportunities—and their return on investment—demands informed, strategic decisions. **Action:** Launch pilot projects and proofs of concept to validate business cases before scaling. Foster cross-functional collaboration to champion informed strategic bets and accelerate value creation.

B2B ENTERPRISE CONDITIONS	VOLATILITY	UNCERTAINTY	COMPLEXITY	AMBIGUITY
Legislative	**Challenge:** Rapidly shifting *laws and regulations* can disrupt critical business operations with little warning, generating operational risk. **Action:** Establish industry-leading *compliance and monitoring systems* supported by dedicated teams. Proactively participate in policy advocacy to influence changes before they impact your business.	**Challenge:** Forecasting how coming *legislative shifts* will affect our operations and strategy is inherently difficult and impacts confidence in long-term plans. **Action:** Deploy *predictive analytics* and rigorous *scenario planning* methods to anticipate, model, and prepare for a range of future legal environments.	**Challenge:** Navigating and enforcing compliance with a *web of regional, national, and industry-specific rules* can fragment processes, drive up costs, and slow decision-making. **Action:** Design *standardized compliance frameworks*—automate processes wherever possible to ensure consistency, reduce errors, and free up strategic resources.	**Challenge:** New or *unclear legislation* often appears, leaving room for interpretation and strategic missteps that can expose the company to unknown risks. **Action:** Engage *legal experts* early and invest in advanced *RegTech* solutions to clarify implications, ensure compliance, and transform uncertainty into strategic clarity.
Financial	**Challenge:** Linked to unpredictable market fluctuations, interest rates, and macroeconomic swings—demanding organizations remain agile and vigilant. **Action:** Proactively deploy financial hedging, real-time analytics, and robust risk management strategies. Prioritize digital transformation to ensure faster, data-driven decisions.	**Challenge:** Forecasting market trends and securing access to capital remains a critical hurdle as economic signals shift rapidly. **Action:** Advance your capabilities by investing in predictive financial modeling, scenario planning, and building relationships with diverse funding sources. Leverage digital tools and AI for forward-looking insight.	**Challenge:** Managing multilayered risks, investment decisions, and increasingly diversified portfolios can overwhelm even experienced teams. **Action:** Establish a comprehensive risk management framework. Employ portfolio management software and automation to create transparency and optimize allocations, supporting resilience even as variables multiply.	**Challenge:** Discerning true signals from noisy or ambiguous financial indicators and economic forecasts in uncertain markets is challenging. **Action:** Empower your teams to synthesize intelligence from multiple data sources. Lean on expert insights and superforecasters to sharpen assessment, enabling faster, more confident navigation through ambiguity.

B2B ENTERPRISE CONDITIONS	VOLATILITY	UNCERTAINTY	COMPLEXITY	AMBIGUITY
Customer	**Challenge:** Rapid shifts in customer preferences and demand can upend even the strongest strategies. **Action:** Implement robust customer feedback systems and leverage AI-driven customer insights to identify new patterns quickly. Staying attuned to these shifts positions your business ahead of market curves and strengthens resilience.	**Challenge:** Accurately predicting future customer behaviors and market needs is more challenging than ever. **Action:** Harness predictive analytics and advanced market research, then develop dynamic customer personas. This anticipatory approach empowers your teams to act decisively and with greater foresight.	**Challenge:** Managing increasingly diverse customer segments demands smarter, more personalized marketing strategies. **Action:** Utilize cutting-edge customer segmentation tools and automation for tailored outreach, ensuring relevance and building trust with each audience. The reward: deeper engagement and lasting customer loyalty.	**Challenge:** Interpreting and acting on evolving customer expectations and feedback can feel ambiguous and disorienting. **Action:** Deploy real-time feedback loops and sentiment analysis to continuously refine your approach. This culture of listening and rapid adaptation transforms ambiguity into actionable insight—and opportunity.
Geographical	**Challenge:** Regional instability, disruptive natural events, and geopolitical tensions constantly reshape our landscape. **Action:** Build a dynamic regional risk management program that anticipates volatility; diversify operations across stable geographies to buffer shocks and seize emerging opportunities.	**Challenge:** Unfamiliarity in new markets and local instability often veil both threats and opportunities. **Action:** Form strategic partnerships with trusted local entities and conduct comprehensive, data-driven regional analyses to gain clarity and shorten your learning curve.	**Challenge:** Navigating complex local regulations, diverse cultures, and intricate logistics stretches enterprise resources. **Action:** Engage respected local advisers and integrate regional compliance management tools to simplify, automate, and standardize risk governance.	**Challenge:** Assessing unknowns in emerging markets can stall decision-making and dilute strategic intent. **Action:** Leverage advanced market entry analyses, real-time risk assessment tools, and scenario planning—transform ambiguity into foresight and competitive advantage.

B2B ENTERPRISE CONDITIONS	VOLATILITY	UNCERTAINTY	COMPLEXITY	AMBIGUITY
Political	**Challenge:** Frequent political upheavals, policy shifts, and trade disputes disrupt the business landscape, creating rapid and unpredictable change. **Action:** Proactively establish robust political risk management frameworks and invest in policy advocacy to buffer the impact of external shocks.	**Challenge:** Uncertainty around evolving regulations, government stability, and future political trajectories clouds decision-making and strategic forecasting. **Action:** Deploy advanced political forecasting models and scenario planning tools to navigate shifting environments with confidence and clarity.	**Challenge:** Navigating diverse and overlapping political systems, regulations, and compliance requirements across regions heightens operational complexity. **Action:** Construct a multitiered government relations strategy that fosters meaningful engagement with authorities at every level, ensuring compliance and influence.	**Challenge:** Political decisions are often ambiguous, making it difficult to decode intent and anticipate consequences for your enterprise. **Action:** Leverage intelligence platforms, risk analysis tools, and expert networks to interpret signals, anticipate moves, and accelerate informed responses.
Climate	**Challenge:** Extreme weather events and rapidly escalating environmental disruptions now occur with unpredictable intensity, creating operational hazards for enterprise leaders. **Action:** Proactively develop climate resilience strategies and embed robust sustainability frameworks across operations. Position your organization not only to withstand shocks but to turn resilience into a strategic advantage.	**Challenge:** The enduring and sometimes irreversible impacts of climate change are difficult to quantify and plan for, increasing planning risk. **Action:** Harness advanced climate impact modeling and foster adaptive scenario planning. Make climate intelligence a core component of future-proofing your business.	**Challenge:** Fast-evolving climate regulations and shifting global sustainability initiatives introduce layers of compliance, reporting, and resource management. **Action:** Deploy enterprise-grade sustainability management systems and drive resource efficiency initiatives that realize measurable outcomes and future-proof compliance.	**Challenge:** Scientific projections on climate risks and environmental impacts are often divergent and difficult to interpret for decision-making. **Action:** Collaborate with climate experts and leverage cutting-edge modeling tools to translate science into actionable business insights. Build capacity to lead through ambiguity and seize opportunity in uncertainty.

B2B ENTERPRISE CONDITIONS	VOLATILITY	UNCERTAINTY	COMPLEXITY	AMBIGUITY
Supply Chain	**Challenge:** Rapid supply chain disruptions caused by geopolitical instability, natural disasters, and volatile market conditions threaten operational continuity. **Action:** Build robust supply chain resilience by diversifying suppliers, investing in adaptive logistics, and deploying real-time monitoring platforms. Anticipate disruptions, and ensure your business responds—not reacts—to crises.	**Challenge:** Ongoing uncertainty makes it difficult to guarantee seamless supply chain performance and resilience when faced with unpredictable events. **Action:** Employ advanced risk management tools and formalize contingency plans to map possible scenarios. Consistently review and update these strategies, ensuring your team is prepared for the unexpected.	**Challenge:** Navigating the tangled web of global supply chains, complex supplier networks, and intricate logistics systems increases risk exposure and demands synchronized operations. **Action:** Leverage integrated supply chain management technologies and cultivate strong supplier partnerships. Actively collaborate across your network to drive efficiency and transparency.	**Challenge:** Ambiguity around emerging risks and their far-reaching effects makes proactive management challenging. **Action:** Harness predictive analytics and scenario planning to illuminate hidden vulnerabilities and clarify potential outcomes. Cultivate an organizational mindset that is prepared for what you can't immediately see.
Sustainability	**Challenge:** Transforming volatility into opportunity requires courage and readiness. **Action:** Lead the change by scaling up sustainability efforts now, positioning your company as a decisive, future-ready leader.	**Challenge:** Uncertainty is the breeding ground of opportunity—if you're proactive. **Action:** Turn ambiguity into bold action with adaptive sustainability strategies that secure your organization's place in tomorrow's market.	**Challenge:** Complexity is merely the sum of untapped advantages. **Action:** Master your operations and inspire your partners, establishing your company as a benchmark for sustainable value creation.	**Challenge:** Where others see ambiguity, you see the space for leadership. **Action:** Set the standard in transparent reporting and stakeholder communication—turn measurement into momentum.

B2B ENTERPRISE CONDITIONS	VOLATILITY	UNCERTAINTY	COMPLEXITY	AMBIGUITY
Innovation	**Challenge:** The pace of technological change is relentless and unpredictable, with new business models and market disruptors emerging constantly. **Action:** Build resilience by institutionalizing continuous innovation. Prioritize investments in R&D and digital capabilities; implement innovation management platforms to rapidly experiment, learn, and adapt. Your competitive advantage depends on embracing change, not fearing it.	**Challenge:** The directions of future breakthroughs are obscured; market shifts frequently upend established best practices. **Action:** Stay ahead of the curve with robust innovation forecasting. Engage in open innovation networks and cross-industry partnerships to access diverse insights. Regular scenario planning and horizon scanning will keep your strategy agile.	**Challenge:** Scaling innovative ideas requires aligning processes, talent, and technology across complex structures, which often leads to friction. **Action:** Drive cross-functional alignment with structured innovation processes and transparent pipelines. Adopt idea management tools to track, evaluate, and accelerate high-potential concepts.	**Challenge:** Strategic value and risks of new initiatives are rarely clear at the outset, making resource allocation difficult and risky. **Action:** Institute rigorous portfolio management and innovation risk assessments. Balance your portfolio with incremental and breakthrough projects to spread risk and maximize returns.

B2B ENTERPRISE CONDITIONS	VOLATILITY	UNCERTAINTY	COMPLEXITY	AMBIGUITY
Growth	**Challenge:** Market conditions, competitive actions, and economic cycles can rapidly shift, impacting all growth projections. **Action:** Leverage robust growth analytics and implement adaptive strategy frameworks that flex with the market. Instill a digital-first mindset, using real-time analytics to detect and respond early to emerging threats and opportunities.	**Challenge:** Identifying sustainable growth opportunities is no longer straightforward, as emerging buyer behaviors and shifting economic drivers create ambiguity in demand. **Action:** Intensify market intelligence and scenario planning. Deploy advanced market research, listen intently to digital buyer signals, and actively seek trends in buyer intent and external influences to sharpen foresight.	**Challenge:** Scaling amid intensified competition, rapid tech adoption, and globalized expectations complicates execution. **Action:** Build scalability into operations. Invest in modular business models and experiment with scalable digital platforms. Recruit talent adept in automation, prioritize process aligns, and continually upskill teams in digital transformation.	**Challenge:** Deciphering growth metrics and forecasting their long-term business impact is more complex as B2B cycles lengthen and channels proliferate. **Action:** Embrace sophisticated growth metrics and dynamic KPIs. Use performance analytics platforms with predictive and prescriptive tools to clarify trends, tie metrics to business outcomes, and align teams to common, data-driven goals.

ACKNOWLEDGMENTS

I would like to thank the many people who directly contributed to the book: Walter Pollard, Brook Spatz, Tim Rawls, Phil Cederstrom, Joe Hayes, Jenny Mittelstaedt, Deb Kammer, Michelle Curtis, our publisher Kory Kirby, as well as our employees, customers, and advocates.

I would like to acknowledge all of the account teams and people that shaped my *Go to Customer* vision.

I would also like to thank our many investors, and in particular our Founding Investor Robert Early as well as Temp Keller, Co-founder and CEO of Templeton Learning.

I would also like to give a special thanks to our partners and influencers who we work with: Bryan Gray, CEO at Revenue Path Group, Brian Shea, Executive Partner at Lucrum Partners, David Brock, CEO at Partners in Excellence, Keith Mescha, Managing Director at EY, Mike Koory, CEO at Blue SalesFly, Ron Hubsher, CEO at Sales Optimization Group, Michael Phelan, CEO at Go to Market Pros, Brynn Tillman, CEO at Social Sales Link, Dan Pfister, Founder at Winback Labs and Dave Lewark, Ted Corbiell, Fred Diamond, Craig Nelson, Keenan, CEO at A Sales Growth Company, Paul Butterfield, CEO Revenue Flywheel Group, and Fred Diamond, Founder for the Institute of Professional Selling as well as Scott Santucci, an original founding member of Polaris I/O.

I would like to acknowledge Gartner Research and the many analysts who contribute to evaluating the future of B2B sales and account planning as well as all those we interface with at the Strategic Account Management Association (SAMA).

Finally, I would like to acknowledge all of the past sales and marketing professionals and analysts who I have worked with over many decades as well as the friends who I have collaborated with along the way.

ABOUT THE AUTHOR

Founder + CEO of Polaris I/O

Dave has 30 years of experience in B2B as President, GM, CMO, and CSO. He is recognized as an AI, marketing, and sales enablement expert. He is also a growth leader with a strong history of innovation.

PolarisIO.com
Linkedin.com/in/irwindave
info@polarisio.com